A Veteran's Journey

Life is a journey, one step followed by another...

A Collection of Short Stories

Volume 1

Andy Adkins

ISBN: 978-1-7363878-2-5 (Paperback)
ISBN: 978-1-7363878-3-2 (Digital Edition)

Published by Andy Adkins, Gainesville, Florida
www.azadkinsiii.com

COVER PHOTO CREDITS:
Cover Design — Andy Adkins
Cover Photos — Andy Adkins

ALSO BY ANDY ADKINS

Never Forget—A Novel

Three Years, Eleven Months, & 29 Days:
But Who's Counting

You Can't Get Much Closer Than This:
Combat with the 80th 'Blue Ridge' Division
in World War II Europe

The Lawyer's Guide to Practice Management System Software

Computerized Case Management:
Choosing and Implementing the Right Software for You

Contents

(1) Beginning a New Journey…

I am a veteran…a United States Navy veteran to be more specific. I enlisted ("volunteered") in July 1973 for four years. I *almost* made four years. It was exactly three years, eleven months, and 29 days (but who's counting).

I joke about that—and I actually wrote a book with that title. But those four years were some of the best. Lots of memories, both good and bad, but you tend to remember the good more than the bad. Isn't that how you successfully get through life?

It's only fitting that I start this new blog today: Veteran's Day 2019. I'm at that age where I can look back at my life & career(s) and wonder, "How did I get here?"

Older people often call that "philosophizing." I call it a journey. I have learned over the years that "life is a journey and you take one step at a time."

Hence, the tagline of my blog.

I grew up in Gainesville, Florida in the 60s & 70s, graduating from Gainesville High School in 1972. To this day, I continue to stay in touch with many of my high school classmates, primarily through Facebook, since we're scattered about. There were more than 1,000 of us who graduated in '72. And there are still several of us in Gainesville.

After high school, I "tried" college (Florida State University), but only lasted a year. I'm not sure why—I was 17 years old when I started in the Fall of '72. After my first year, FSU asked me to leave. I believe the term on my report card was "Dismissed for academic reasons." I maintained a 0.7 GPA. Why would they want me to leave?

Then I joined the Navy.

We were still fighting the Vietnam War, though the draft officially ended in January 1973 (a few months before I joined). But since I didn't know what I wanted to do with the rest of my life, I thought I'd give the military a try.

My father served in the U.S. Army, 80th Infantry Division, and fought the Germans during WWII (I wrote a book about that, too). I gave that a quick thought. But at the time, I was into scuba diving. One of my dive buddies had served in the Navy and had good things to say. My thought process was simple: there were probably more scuba diving opportunities in the Navy than the Army. There, I made a decision.

Since I was 18, I was able to legally sign documents. I rode my bike downtown to the Tallahassee Navy Recruiter's office and signed up for a four-year enlistment.

From my Navy book: *A few days later, I was working at the dive shop when my mom and dad called. My dad said, "Son, we just got your report card and wanted to talk with you about your grades." "Yeah, I know. I didn't do as well as I should have." Before my dad or mom could say another thing I blurted out, "I joined the Navy." There was a silence—a long, deafening silence. I think I heard a collective gasp from the other end of the phone. My mom said, "What?" I told them again.*

I forgot whatever else was said during that phone call, but my mom told me years later that she and my dad were very proud of me at that moment. They said that I had made one of the most important decisions in my life. More importantly, I stuck with it, all three years, eleven months, and 29 days.

I had a great, short Navy "career." After boot camp in Orlando, I went to Lakehurst, NJ to attend ABH (Aviation Boatswain's Mate-Handler) "A" school. That was my specialty: aircraft crash & rescue firefighter and aircraft director.

My first duty station was NAS (Naval Air Station) Agana, Guam. That's on the other side of the world, by the way. But lots of great scuba diving.

After 15 months in Guam, I transferred to an aircraft carrier, *USS Kitty Hawk* (CV-63). I spent two and a half years aboard *Kitty Hawk*. We made one six-month Westpac cruise in 1975, then headed up to Bremerton, WA for a year in the shipyards.

All-in-all, I figure I had three "tours of duty" in four years. *One step follows another.*

After the Navy, I returned home to Gainesville and took a job as a firefighter with the City of Gainesville. In 1977, I was making $9,600 a year. To me, that was a lot of money back then.

I'll spare you the details, but I went back to school (University of Florida this time), got a couple of degrees in Electronics Engineering, held several engineering jobs in different locations throughout the country, and eventually started my own business as an independent legal technology consultant.

Long story short, it's been quite a journey, *one step followed by another.*

Now, I'm in my third retirement phase, or whatever you call it, and I'm writing again. I've written several books, both business (legal technology) and military non-fiction (WWII and Navy). As a consultant, I wrote and published more than 300 articles. Part of that was for marketing. But I like to write. It's become addictive.

And I just completed my first novel. While I'm waiting for the right literary agent and the right publisher (the "Query Trenches"), I'm beginning to work on my next novel. I mean, why wait.

So, that's my backstory.

(2) Following My Father's WWII Path

This year marks the 75th anniversary of the liberation of Europe from terror. For us baby boomers and veterans, this is a somber reminder of what our fathers and grandfathers sacrificed in order to preserve the freedoms we've come to enjoy today.

My father, A.Z. Adkins, Jr., fought with General George Patton's Third Army during the war. He was with the 80th Infantry Division, which landed on Utah Beach in early August 1944. He was wounded in early October in Sivry, France, sent to England for rehab, and then returned to the 80th the day before the Battle of the Bulge began. He was also one of many who "stumbled upon" several German concentration camps, most notably, Buchenwald.

I can't tell you the chills I had when my wife & I visited Europe (France, Luxembourg, & Germany) a few years ago. In several cities, the local citizens celebrated their liberation by the 80th Division by unveiling several memorials.

Knowing that I was standing in the same locations my father had stood years earlier was an honor. Trekking deep into a forest with a local historian, I stood in a foxhole where my father *may* have slept.

But I know his company (Company H, 317th Infantry Regiment) launched mortars from that location in order to fight off the Germans. That was an extraordinary journey for me. It still gives me goosebumps when I think of that special day.

I've been on two of these trips, following my father's journey through the war.

My wife's father, Dr. Rufus K. Broadaway, was also in WWII; he parachuted in behind enemy lines during the early morning hours on June 6, 1944—D-Day. He was with the 82nd Airborne Division.

We took the opportunity to visit the Normandy area, including Omaha Beach and Utah Beach. Being able to follow in our fathers' footsteps, so to speak, was a life-changing experience for us both. The French, Luxembourgers, and Germans are perfect hosts and at every turn, they showed their appreciation for what our veterans did for them.

We also visited the Luxembourg American Cemetery, located in Hamm, Luxembourg, just outside Luxembourg City. My wife and I were with many WWII veterans and their families and joined Helen Patton, General Patton's granddaughter, in a moving ceremony. We strolled around and through the well-kept cemetery grounds; again, a very emotional experience—there are more than 5,000 American war dead buried at the cemetery, which covers 50 acres and was dedicated in 1960.

A friend of mine, also on the tour, visited his older brother whom he lost in WWII. A photo I took of him at his brother's grave is one of the most moving photographs I've taken and a constant reminder of what war does to a family.

There will be many celebrations over the next few months in Europe. Learn what you can about your fathers and your grandfathers and what they did during WWII. And if you have the opportunity, I hope you'll take advantage of some of these activities and celebrations the French, Luxembourgers, and Germans are planning for the 75th Anniversary of their liberation.

(3) Between a Rock and a Hard Place

You've seen movies with aircraft carriers, haven't you? *Top Gun* (1986), *The Final Countdown* (1980), *Behind Enemy Lines* (2001), to name a few. You probably wondered what it's like to work on the flight deck, too. Well, maybe not everybody, but stay with me on this.

Most of the time, movies only show one or two aircraft launching. After all, that's where the action is. But in reality—at least in the mid-70s when I was working on the flight deck of *Kitty Hawk*—we'd launch 20-25 aircraft, one right after the other during a normal flight operations cycle.

Once those birds were in the air, we'd move aircraft around (called a "respot") to get ready for the next launch cycle. A typical launch cycle on a good, clear day may take 10-15 minutes to launch all 20-25 aircraft.

When the second launch is completed, we'd have to recover that first set of aircraft back on board. That's fun, too. Once again, we'd "respot" the remaining aircraft to get them out of the landing area in order to recover the others. Typically, we'd move them to the bow (front) and to the starboard (right) side of the ship. There's not a lot of time between flight cycles. While some of the birds may be able to refuel in mid-air, not all do.

Modern day aircraft carriers are built with the landing area at an "angle" to the mainline of the ship. I'll spare you the details, but you can imagine the difficulty of trying to land an aircraft at 80-100 mph at a 12-14 degree angle while the ship continues to move in a forward direction, always heading into the wind.

At the end of the angle in what is commonly called the "crotch," there stands an aircraft director (ABH—Aviation Boatswain's Mate-Handler), called a "Yellow Shirt," because he (or s/he now) wears a yellow jersey. That was one of my jobs.

This particular individual's responsibility is to direct the recovered aircraft out of the landing area quickly and safely so the next aircraft can land. On a good day, aircraft land on the flight deck about one per minute. Not much time for mistakes.

In my opinion, this was *the* most exciting job on the flight deck. I was often told to "think about where you are and where you want to jump."

The job was simple: wait until the aircraft landed safely, catching the arresting gear cable with its tail hook, run out onto the flight deck to meet the plane, making sure the pilot sees you (you're about twenty feet away), and give him a hand signal to raise his tail hook in order to retract the arresting gear cable.

If the hook raised properly, direct him to taxi out of the recovery area, turning him to the starboard (right) side, give him the hand signal to fold wings, and pass him along to another Yellow Shirt director. Then thirty seconds later, do it all over again. *Easy, peasy,* right?

On one particular day, we were running flight ops as usual. I was in the crotch directing aircraft out of the recovery area. It was a "Champagne Day," as my mother used to say—gorgeous, clear and crisp.

During this particular flight op, we were launching aircraft from the bow catapults at the same time as recovering. We had recovered the first batch of F-14 Phantoms and A-6 Intruders and were bringing aboard the A-7 Corsairs. Simultaneously, we were launching F-14s off cat #2, right next to me. It's extremely loud, even with two sets of ear protection, when you are that close to an F-14 in full afterburner.

After the third A-7 Corsair landed and I taxied him out of the recovery area, I was watching the next one approach. He was flying a little awkward. I naturally assumed he was a new pilot still working on his quals and not used to landing on a pitching flight deck. Meanwhile, an F-14 was preparing for launch right next to me. My attention was bouncing between the F-14 in full afterburner and the approaching A-7. I was squatting down, ready to spring into action when the A-7 landed.

As the A-7 hit the deck, it grabbed the arresting gear cable with its tail hook and the pilot applied full power—normal procedure. They do this in case the tail hook misses the arresting gear cable and they need to keep flying off the angle to return for another approach. The only problem was that this A-7 pilot not only applied full power, but also pulled back on his elevator at the same time. He had the cable in his tail hook, yet he was also in the air fixin' to make a big splat.

Let me pause here at this particular moment in time and relay to you what was going on. For one, the A-7 Corsair was in the air about ten feet above the flight deck with an arresting cable in his tail hook, soon to splatter on the flight deck. I was less than fifty feet away. To my left, about twenty feet away, was an F-14 Tomcat in full afterburner, ready to launch. To my right, about five feet away, was a sixty foot drop down to Davy Jones' locker. It was not looking like a good day. I basically had two choices: one, become instant toast or two, find out if Davy Jones really does have a locker.

At that moment, I think *every* flight deck crash training film I had ever seen flashed through my mind. I'd heard of an A-7 Corsair earlier on the *USS Constellation* that had done the exact same thing. But when he slammed down onto the flight deck, the nose gear pierced the cockpit, right between the pilot's legs. The pilot was OK, but the A-7 was not. No fire, thank God, but it took a while to clear the recovery area. That was the thought that was currently in my mind.

I quickly jumped up. The Catapult Officer ("Shooter") on cat #2 had seen what was happening and immediately went into an abort mode, shutting down the F-14. The A-7 slammed onto the deck, but the nose gear held and did not go through the cockpit. The pilot was a little shaken, as was I. But we both managed to get the A-7 out of the recovery area without having to change skivvies. I was fine after that, but it did shake me up a little bit. My good friend and fellow Yellow Shirt, Smitty, came over and asked if I needed a break; I took him up on his offer. My dad told me later when I relayed this particular story, "Builds character, doesn't it?"

Yes, the Navy did help mold my character, not only that particular day, but throughout my life. *Teamwork* is the key word here. The average age of sailors working the on flight deck back then was 18-19 years old. To work on a flight deck with a couple of hundred other guys, most of whom you did not personally know, you had to keep your head on a constant swivel as well as watch out for others.

Working on the flight deck of an aircraft carrier has often been called the most dangerous job in the world. I wouldn't argue that. It was just part of my journey.

(4) From Civilian to US Navy

I was 18 years old when I joined the Navy. I wrote about the "why" in an earlier blog. I already had a year of college under my belt (FSU). Even though I flunked out ("dismissed for academic reasons"), I was eligible to join as an E-2—that was the second-level up in pay grade for an enlisted man.

The Navy recruiter told me to report to the Jacksonville, Florida Induction Center (now called the Military Entrance Processing Command) on Tuesday morning, July 10, 1973 at 7:00 a.m. "sharp." To make sure we arrived on time, the Navy was kind enough to put us new recruits up in the Heart of Jacksonville Hotel the night before.

At five a.m. (aka, "0'dark hundred"), all potential recruits received a *not-so-gentle* wake-up call. We herded down to breakfast at the hotel, checked out, and ushered onto a bus that would take us to the Induction Center. For most of us, it would be our last ride as civilians.

I don't remember all the wonderful things we did during the day, other than waiting around ("Hurry up and wait"), talking with Navy personnel in their sparkling-white summer uniforms, and filling out endless reams of paperwork. This *was* the early 70s; there were no computers.

Has anyone ever told you to "Turn your head and cough"? Stay with me here.

You hear that phrase quite a bit in the Induction Center. During your physical, you stand in a line in your underwear along with a couple of dozen other recruit wannabes. All are facing forward, standing at near-perfect attention.

I say "near-perfect" because we didn't know what "stand at attention" meant at the time. We just mimicked what we remembered from the movies.

The doctor comes in with a Navy hospital corpsman following close behind. The corpsman carries a box of latex gloves—this does *not* look good.

"You men on the right, turn your head to the right. You men on the left, turn your head to the left. Now drop your shorts."

I'll spare you the details, but needless to say, it was a very *personal* lesson in humility. God only knows what he was looking for. I certainly wasn't hiding anything. I thought to myself, "What does this guy do on vacation for fun?"

After losing all sense of privacy, those of us who passed the physical exams (I often wondered who couldn't pass *those* tests) were sworn in. While it was a pronounced speech, I felt that I had taken a major step in my journey called life. I was now enlisted in the United States Navy.

We left the Jacksonville Induction Center later that day on a Trailways bus heading south to Orlando, about a 3-hour drive. We arrived at the Orlando Naval Recruit Training Center (RTC) around midnight and were met by a big, burly guy dressed in Navy whites.

We were placed in a temporary holding company for a few days, waiting until we had enough recruits to build a company of eighty men.

We were a mixture of all sorts: guys who had nowhere else to go, guys who had the choice between going to jail or joining Uncle Sam's Sailing Club, and guys like me who wanted to become a man. In other words, most of us didn't have a clue what we wanted to do with our lives.

Before I continue, let me pause here for a brief moment to relate something very personal to you. I grew up in a fairly conservative home. My father was a WWII veteran, a lawyer, *and* a Methodist. My mother was one of four sisters and grew up in a Baptist household. Both my parents were born and raised in Florida. I grew up attending the First United Methodist Church. Other than a few choice swear words I learned in high school, we simply didn't use curse words while I was growing up.

The first morning in our new recruit company (Orlando RTC Company #163) we woke to the sound of a fifty-gallon metal garbage can being kicked down the middle of the barracks and some big burly guy shouting, "Get the f*** up. Drop your c***s and grab your socks. You've got fifteen minutes to s***, shower, and shave."

Welcome to the United States Navy—it's 0400 (pronounced "*O four hundred*"). That's four o'clock in the frigging morning. Before heading to the bathroom, we first had to stand in front of our bunks to be counted to make sure no one had jumped the fence to go AWOL (Absent without Leave). Fortunately, no one fled *that* night.

Chow was served in the mess hall, about a half-mile away from our barracks, starting at 0600 (6:00 a.m.). Chow lines were long, but they moved fast. Here, I was introduced to one of the most beloved meals in the Navy: S*** on a Shingle—chipped beef in white gravy on toast.

It wasn't half bad, especially if you added enough salt and pepper. But I was wondering, where were the bacon and eggs and pancakes my Navy recruiter promised? Where were the biscuits like my mom used to make? My God, how can anyone eat this crap?

The first couple of days we were lined up for all sorts of fun activities, starting with haircuts. By lined up, I mean we were nice and tight, or as you get more used to it, the Navy term is "Nut to Butt." For those of us who grew up in the seventies, long hair was "in" and most new recruits had long hair.

We were all lined up, standing at attention outside the barber shop. By this time, we knew what "stand at attention" meant. There were six barber chairs in the shop. There was also about a foot of hair on the floor, almost like walking through a field of wheat. I hopped into the chair and while my hair was not that long, I asked the Navy barber to "take a little off the top." Well, after thirty seconds (the barbers had a contest to see who could shave a new recruit's hair off in the shortest amount of time; like shearing sheep), he was done with me and said, "How's that smartass? Now, get the f*** out of my chair." Welcome to the United States Navy.

Besides being woken up at 0400 with some lard ass kicking a garbage can down the middle of the barracks and screaming vulgar obscenities at the top of his lungs, we got into a somewhat regular routine those first few days: early morning calisthenics on the "Grinder," three square meals a day, learning to march, and cleaning our barracks. I was even learning a new language—"Navy lingo."

I thought to myself, this isn't so bad. Boy, was I in for a surprise.

(5) My Current "Journey"
A Novel Yet to be Published…

Over the span of my career(s), I've published more than 300 articles, made more than 250 presentations, and written several books. I like to write and most of the time, I like to think I have something worthwhile to say.

It started out as a "I need to write" for my consulting business: it was a necessity and definitely good for marketing my legal technology expertise. It gradually morphed into a "I need to write" because I love to write. It's become a passion.

Like most writers & authors (I use those terms interchangeably), I know my writing has improved over the years. Periodically, I'll read an old article I wrote back in the late 80s / early 90s and realize not only how technology has changed (I'm a computer guy), but also my writing style has changed.

I know many of my lawyer friends can relate: even though "legal drafting" was a required course in law school, those first few "real-life" legal documents were not as honed as their current writing style.

And no, that's not just technology changes. Face it, it's a sign of experience and more important, a sign of maturity.

Just in case you're interested (you know you are), I published "What I've Learned in the Last 20 Years" in 2008. This article in itself is a journey from DOS to Windows, from WordPerfect to Microsoft Word, and into the Internet and how the legal profession has changed over the course of "tech time." Much of what I wrote eleven years ago is still prevalent in today's wireless techno-world.

Since I'm "retired," I needed another challenge. I turned my attention to writing another book. I mentioned earlier I've written several books. Here's the list:

- *WordPerfect Law Office Solutions for DOS* (2 chapters)
- *WordPerfect Law Office Solutions for Windows* (2 chapters)
    - Published by James Publishing, 1995 (out of print)
- *Computerized Case Management Systems, Choosing and Using the Right Software for You*
    - Published by the American Bar Association, 1998 (currently out of print)
- *You Can't Get Much Closer Than This: Combat with the 80th "Blue Ridge" Division in WWII Europe*
    - Published by Casemate Publishing, 2005
    - Selected as the "Book of the Month" for the Military Book Club in November 2005
- *The Lawyer's Guide to Practice Management Systems Software*
    - Published by the American Bar Association, 2009
- *Three Years, Eleven Months, 29 Days—But Who's Counting*
    - Published by AZAdkinsIII, 2014

Which brings me to my latest and perhaps my greatest, **NEVER FORGET**. This is my *first* novel and I'm quite proud of it (who wouldn't be?). I've completed the 94,000-word manuscript and searching for a literary agent to take on this project. While I'm currently in the "Query Trenches," I'm continuing to write.

A word about my new book. **NEVER FORGET** is a father & son story of overcoming personal conflicts caused by war. In one sentence: *It's the journey of a Vietnam veteran to find redemption and forgiveness.*

I use the following "pitch" in my Query Letters I send to literary agents. From what I understand, this blurb should read similar to the description you'd find on the inside flap of the book cover.

NEVER FORGET *by Andy Adkins*

In 2001, Vietnam War veteran Tom Reilly lives with a haunted past he has tried to forget. His father, Ed, has recently been diagnosed with Alzheimer's. Though they have not spoken for years, Tom visits Ed at a retirement community for the first time in 25 years—the fighting continues, but the healing begins.

Tom has tried to live a simple life. After the tragic death of his wife years earlier, he is left to raise his son alone. Chris has never met his grandfather. Yet, when the three of them meet, Chris and Ed form a unique bond, helping Tom better understand his own father. Chris interviews Ed for a high school history project and, while listening to his father answer questions about WWII, Tom internalizes his own answers about Vietnam. They dare to compare their own wars and hear each other's combat stories for the first time.

Ed knows there isn't much time left to repair what was broken between them. With the help of a resident psychologist and an aide in the memory unit, both with deep, personal ties to Vietnam, Ed encourages Tom to begin the difficult journey of healing and forgiveness. The question is, can they resolve their differences before it's too late—for both of them?

NEVER FORGET is a multi-generational father, son, and grandson novel about WWII and Vietnam. It is *not* about combat; it is about overcoming personal conflicts. The story takes place in 2001 and reflects upon our current society of an older generation (WWII), aging baby boomers (Vietnam), and the current generation (global terror). It brings together multiple facets of dementia and survivor's guilt in a moving story of personal struggles and growing relationships.

(6) Military Lingo

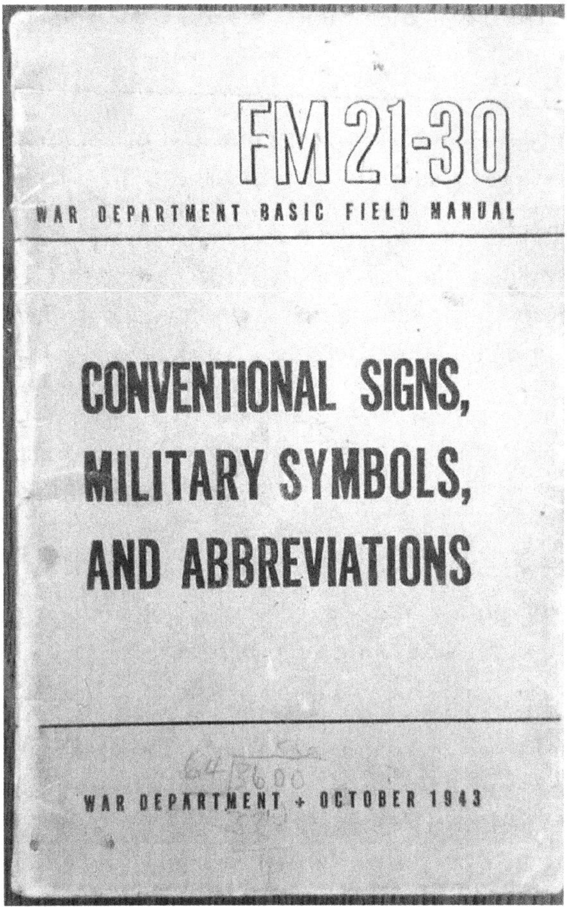

Note to self/readers: No "caution" is needed for *this* particular post.

However, in the "linked" pages that may (or may not) appear later in this post, the reader [Pay attention: this is the legal "disclaimer"] should be aware that these glossaries include three languages: English, military, and Navy (or Army/USMC).

The reader may find two of the three offensive.

I mention the above statement tongue-in-cheek, but our men and women in the military have a certain language (or "lingo") and a unique way of speaking their mind.

Oh, you didn't know that?

However, once we leave the service, much of that colorful language often continues to be a part of our daily lives.

With that tidbit of knowledge in mind, let me introduce you to a bit of the *cleaner* military "lingo." Note that the Navy, Army, USMC, and USAF use many of the same terms and phrases, but since *yours truly* is a Navy veteran, I tend to use "Navy speak" more than the others.

Here are some of my favorites. I introduced my wife to these terms early in our marriage; my kids are also somewhat familiar with the lingo. No, I have not taught this language to my grandkids (yet).

AWOL—"Absent Without Leave."

Beat Feet—a phrase used to describe someone getting away quickly. "Let's beat feet outta here before the Chief of the Boat comes back."

Boats—a shortened nickname used to describe a sailor in the Boatswain's Mate rating. Also, a term often used by veterans to call their children (or maybe that's just me). "What's up, Boats?"

Boonies—a slang term used to describe remote, wild areas, jungles, or swampy areas.

Boot—a slang term for a new recruit. Also known as a "Cherry" or "New Meat."

Brain Fart—a phrase used to describe a condition when, under stress, one cannot recall or perform something that would normally be easy or second nature. "Did you just have a brain fart when you walked into my office?"

Bummer—a term to describe bad luck. "The Gators lost again; bummer."

CF—"Charlie Foxtrot." I'll just leave this for you to Google.

Chow—another term for a meal. "What time is chow?"

Cop out—a phrase used to describe when someone quits or backs out of an assignment.

Days and a Wake Up—a phrase used by short timers to describe how much time left before discharge or a transfer. "I've got 10 days and a wake up."

Decked out—a phrase used to describe someone who is all dressed up. "He's really decked out for tonight's party."

Drag—a term used to describe a situation that is disappointing. "The Gators are losing again; that's a drag."

Five-finger discount—a phrase used to describe the cost for something that is acquired by theft. "How much did that cost you?" "Nothin. I got it with a five-finger discount."

Gedunk—refers to snack food, such as candy or chips, as well as the place they are sold: The Gedunk Shop.

Grunt—a U.S. infantryman, also known as a "Ground pounder."

Gung ho—a phrase used to describe someone who is enthusiastic and ready to go. "He was all gung ho when he joined the Navy."

Head—another term for a bathroom.

Hurry Up and Wait—a phrase used across all military branches to sarcastically describe the military's propensity to perform tasks quickly, then sit idly by for long periods of time.

KP—"Kitchen Police." A term used to describe duties in the galley, usually NOT cooking, but cleaning. "I can't play poker tonight, guys. I've got KP."

Lay it on me—a phrase meaning to speak your piece, tell me what's on your mind. "I need someone to talk to." "I'm here; lay it on me."

Living the dream—a sarcastic phrase used when someone is asked how they are, they reply with this which sounds upbeat and a positive term, and they are actually miserable. "How are you doing today Mr. Jones?" "Living the dream, Sir!"

Make a Hole—a phrase used by all military branches to "suggest" to a group of people to get out of the way. In boot camp, we'd always say (out loud, but not shouting), "Make a hole, coming through."

Marine Candy Bar—a derogatory phrase used by the Navy to describe a urinal cake. I can't see Marines using this term, unless it's to describe a "Navy Candy Bar."

Mid-rats—short for "MIDnight RATions." On a ship, usually leftover lunch and dinner or a PB & J. Served after hours, around midnight.

NAVY—"Never Again Volunteer Yourself."

Non-skid—a rough epoxy coating used for grip on a ship's weather decks (including the flight deck of an aircraft carrier). Also, it's another term for toilet paper. "Honey; can you pick up a package of non-skid when you go to the store?"

Nut to butt—a phrase used to describe standing in line—usually at attention—close quarters, body to body, each man's chest pressed to the back of the man ahead, or "nut to butt." Similar, but not quite, to riding the subway or London tube at rush hour.

O dark thirty ("Oh Dark Thirty")—a phrase used to describe a time before pre-dawn, very early in the morning. Also, "Zero-dark thirty."

Occifer—a derogatory term used to describe an officer.

Polish a turd—a phrase used to describe how to make the most of a bad situation.

Port—the left side of the boat or ship (when facing the bow). It is also the left side of an aircraft when facing the nose from inside. Also a term for place of arrival for ships.

Pucker factor—a phrase used to describe the tension caused by high stress during a difficult or dangerous evolution. So named because your sphincter tends to tighten up, or "pucker" involuntarily, during such times. If you've ever been to the Chicago Sears Tower (now called the Willis Tower) and walked onto the Skydeck… on the 103rd floor… and looked down… the "pucker factor" is about a 9.0.

Scuttlebutt—a term used to describe rumors; also a term for a water fountain. "What's the latest scuttlebutt?"

Skivvies—another term for underwear.

SOP— "Standard Operating Procedure."

Starboard—the right side of the boat or ship (when facing the bow). It is also the right side of an aircraft when facing the nose from the inside.

US ARMY— "Uncle Sam Ain't Released Me Yet."

USMC— "Uncle Sam's Misguided Children."

WTF— "Whiskey Tango Foxtrot." Like "CF" (above), I'm just going to leave this one right here.

Now you know some of the more common. **Navy Lingo and Army/Marine Lingo** terms and phrases.

You are welcome.

(7) My World War II "Journey"

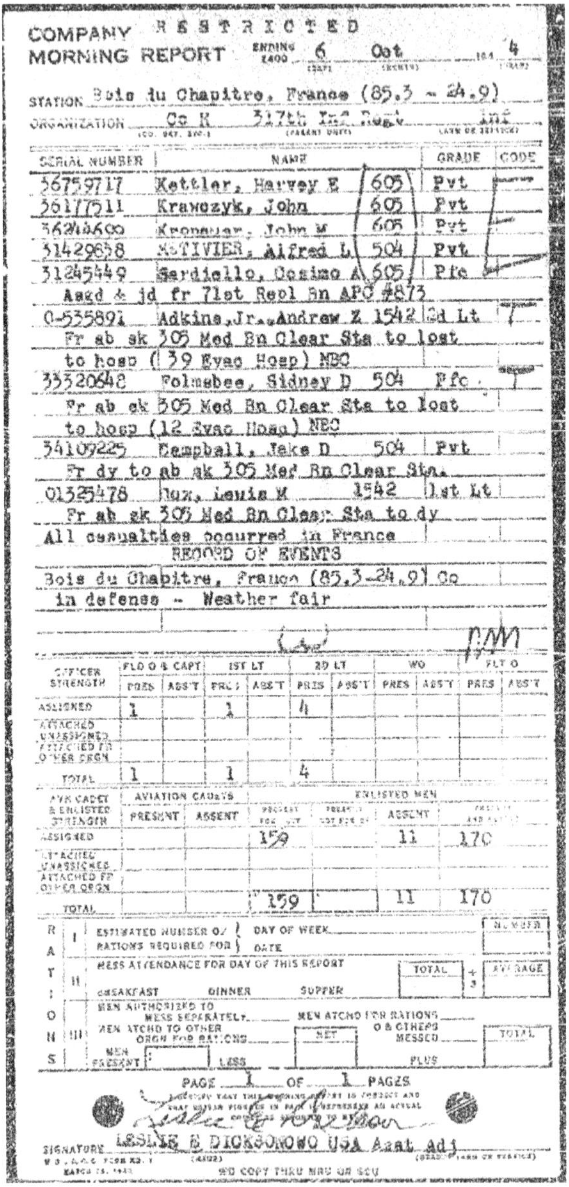

COMPANY MORNING REPORT RESTRICTED

ENDING 2400 6 Oct 1944

STATION Bois du Chapitre, France (85.3 - 24.9)

ORGANIZATION Co K 317th Inf Regt Inf

SERIAL NUMBER	NAME		GRADE	CODE
36759717	Kettler, Harvey E	605	Pvt	
36177511	Krawczyk, John	605	Pvt	
36244600	Kronauer, John M	605	Pvt	
31429658	METIVIER, Alfred L	504	Pvt	
31245449	Sardiello, Cosimo A	605	Pfc	
Asgd & jd fr 71st Repl Bn APO #873				
0-535891	Adkins, Jr., Andrew Z	1542	2d Lt	
Fr ab sk 305 Med Bn Clear Sta to lost to hosp (39 Evac Hosp) NBC				
33320648	Folmsbee, Sidney D	504	Pfc	
Fr ab sk 305 Med Bn Clear Sta to lost to hosp (12 Evac Hosp) NBC				
34109225	Campbell, Jake D	504	Pvt	
Fr dy to ab sk 305 Med Bn Clear Sta.				
01325478	Nox, Lewis M	1542	1st Lt	
Fr ab sk 305 Med Bn Clear Sta to dy				
All casualties occurred in France				

RECORD OF EVENTS

Bois du Chapitre, France (85.3-24.9) Co in defense - Weather fair

OFFICER STRENGTH	FLD O & CAPT PRES	FLD O & CAPT ABS'T	1ST LT PRES	1ST LT ABS'T	2D LT PRES	2D LT ABS'T	WO PRES	WO ABS'T	FLT O PRES	FLT O ABS'T
ASSIGNED	1		1		4					
ATTACHED UNASSIGNED										
ATTACHED FR OTHER ORGN										
TOTAL	1		1		4					

AVN CADET & ENLISTED STRENGTH	AVIATION CADETS PRESENT	AVIATION CADETS ASSENT	ENLISTED MEN PRESENT FOR DUTY	ENLISTED MEN PRESENT NOT FOR DUTY	ENLISTED MEN ABSENT	ENLISTED MEN PRESENT AND ABSENT
ASSIGNED			159		11	170
ATTACHED UNASSIGNED						
ATTACHED FR OTHER ORGN						
TOTAL			159		11	170

RATIONS I ESTIMATED NUMBER OF RATIONS REQUIRED FOR DAY OF WEEK DATE NUMBER

II MESS ATTENDANCE FOR DAY OF THIS REPORT BREAKFAST DINNER SUPPER TOTAL ÷ 3 AVERAGE

III MEN AUTHORIZED TO MESS SEPARATELY MEN ATCHD TO OTHER ORGN FOR RATIONS MEN PRESENT LESS NET MEN ATCHD FOR RATIONS O & OTHERS MESSED PLUS TOTAL

PAGE 1 OF 1 PAGES

I CERTIFY THAT THIS MORNING REPORT IS CORRECT AND THAT WITHIN FIGURES IN PART I REPRESENT AN ACTUAL COUNT

SIGNATURE LESLIE E DICKSON 0 USA Asst Adj

W D, A G O FORM NO. 1 MARCH 15, 1943

WD COPY THRU MRU OR SCU

For Christmas 1984, my dad gave me and my sister, Anne, a 200-page, three-ring binder titled, **A World War II Diary.** I had no clue that he had kept a diary of his time during the war, so this was a total surprise.

Of course, I knew he had fought in WWII in Patton's Third Army with the 80th Infantry Division, but little else. Here I was, in 1984, holding a day-by-day account of the most tumultuous time of my father's life.

And so began *my* World War II "journey."

Like many baby boomers—sons and daughters of World War II veterans—I grew up hearing occasional "war stories" and learned enough to know what a good job our soldiers had done. But like so many other veterans, my father simply did not talk about the *real war:* the blood, the death, and what hurt the most—the loss of his men and his Army buddies.

I read his diary within a few days. I reread it again and again. Each time, I learned something new about my father. It was hard for me, a child of the 60s, to fathom what he had been through, what he had witnessed, and what he had lost.

But there it was, right in front of me, in black and white.

Well… I got busy with… you know… life.

While I was able to share a few stories with my father, I was never able to make the time to really talk with him about the war—more than what was in the diary. To tell you the truth, I'm not sure he wanted to talk about it. I think writing this diary helped bring closure and put the war behind him.

My father died in June 1989, a few days shy of his 68th birthday.

A year later, I scanned his diary into the computer (I *am* a computer guy). This was way back in the days of DOS 3.3 and WordPerfect 5.1 (those *were* the days). Since it was a diary, a day-to-day journal, I wanted to organize it (chapters), expand on it (research), and possibly publish it.

Little did I know what I was getting myself into at the time.

My initial thought was to get the diary/book into a form that I could pass along to my children, family & friends, and other 80th Division veterans. When I first seriously started to consider publishing the book, I found only a handful of WWII memoirs. I don't know why, but perhaps it's because "The Greatest Generation" didn't talk much about their war.

Several years later, in October 2005, Casemate Publishers published, ***You Can't Get Much Closer Than This: Combat with Company H, 317th Infantry Regiment, 80th Division***, by A. Z. Adkins, Jr. and Andrew Z. Adkins, III. It was immediately selected as the "Book of the Month" for the Military Book Club in November 2005.

I was a proud son.

When those first books arrived at my front door, I literally cried. I inscribed the first one to my father and took it out to the cemetery to share with him. If nothing else, that simple gesture (I have tears flowing as I write this, even after 14 years) brought me closer to my father.

Stay with me. The journey hasn't ended.

Since that time, I became more involved with the 80th Division Veterans Association, my dad's old Army division. I created the 80th Division website (www.80thdivision.com), digitized more than 100,000 pages of WWII Morning Reports, General Orders, Unit Histories, After Action Reports, and dozens of Miscellaneous Reports.

All are free and available for download from the website. I also became the association's National Secretary for several years and was awarded, "Honorary National Chairman."

My initial intent to get more involved with the 80th Division was simple: I wanted to know more about my father. Many times over the years, I have been humbled by meeting some of the 80th men he fought alongside with during the war. Unfortunately, most of them have since passed away. But at the time, I was sure other veterans' children and family members would want to know about their fathers and grandfathers, too.

Little did I know what I was getting myself into.

I also know that at least six books have been written and published by other 80th Division veteran family members with information gleamed from the documents I've placed online.

This is why I do what I do.

Even today, I continue to receive about 4-5 emails a month from family members seeking additional information about their 80th Division veterans.

In most cases, I can point them to specific reports. I *personally* know from my own experience that when they see their father's (or grandfather's or uncle's) name on a Morning Report, it has an immediate emotional impact.

I cannot tell you how many gracious emails I've received from these family members. This is such a simple task for me and I am honored and pleased to be able to help.

I am thankful that so many people want to continue the legacy of these courageous men. I know other WWII historians who also recognize the importance of our mission.

This is one of those continuing rewards that *I never expected* when I first published the WWII book in 2005.

I've read some of these emails to my sister and to my wife to let them know, "This is why I do what I do."

I'll share with you an 80th Division veteran family member request.

Several years ago, I was contacted by an individual who wanted to know anything I could find about his uncle who had served with the 80th Division. He provided a little information about him (you always need to know what unit) and I began my research.

Normally, a soldier appears in one, two, or three Morning Reports (MR). For example, if a soldier is WIA (Wounded in Action), his name would appear on the daily MR when he was wounded. His name may appear a few days later if he RTD (Returned to Duty), or if he DOW (Died of Wounds). That's how the Army company clerk reported the unit's daily activities.

In this one particular case, I found the soldier who, on the first report, was a Corporal, but went AWOL (the MR never states "why"). He was picked up a few days later (the next MR) and returned to his unit. A few days later, he was busted (demoted) to Pfc (Private First Class).

A week or so later, he went AWOL, *again.* Then returned. Then busted to Pvt (Private). This went on for several more reports over the course of at least a month. All in all, I believe there were more than a dozen MRs where his name appeared—going and coming, so to speak.

I doubled checked to make sure I was reading the Morning Reports correctly. I had the right ASN (Army Serial Number) on all the reports.

How do you inform someone that their veteran family member went AWOL and was busted multiple times? The only approach I could take, which I learned was the best approach, was to NOT try to analyze what happened and why, but to simply report the facts. Which I did.

The family member sent me a gracious email. Something along the lines of: "Well, that was Uncle Jack. Always a troublemaker, but he was a good man with a big heart. Thank you for helping us better understand why Uncle Jack was the way he was."

And that is why I do what I do.

(8) My Dad's "Journey": Battle of the Bulge, Christmas 1944

Like many of my friends, classmates, and veterans, my father fought in World War II in Europe. He was an officer in the 80th Infantry Division and landed on Utah Beach on August 5, 1944. He was a "mortar man," in charge of a "section" of .81mm mortars.

He was wounded in October near Sivry, France. An enemy artillery shell hit the house where he was directing fire as a forward observer for his mortar teams.

The roof caved in, knocking him out, and pinned him down. It also tore up his knee. He was sent back down the line and ended up in England for two months of rehab after knee surgery.

He returned to his unit on December 14, 1944—two days before the Battle of the Ardennes (aka, Battle of the Bulge) began—75 years ago.

This blog article is dedicated to my dad and the thousands of courageous men who fought and endured that long, cold & bitter winter under the worst possible conditions one could imagine.

# # #

My first Christmas in Guam (1973) was my first Christmas away from home. I was in a beautiful island paradise… warm breezes… and three square meals a day. I had just turned 19 years old.

My dad's first Christmas away from home was spent in Luxembourg… in sub-zero temperatures… out in the open… getting shot at and shelled by German artillery. My dad was 23 years old.

The following are excerpts from my WWII book, ***You Can't Get Much Closer Than This: Combat with the 80th 'Blue Ridge' Division in World War II Europe.***

I returned to my unit after two months in England in a hospital recovering from a wound. After I had been back a couple of days, German Field Marshall von Rundstedt launched his big counteroffensive, now known as the Battle of the Bulge.

The weather was bad—snow, freezing cold, and fog. Under this cover, the Germans moved three field armies and surprised the Allies, overran them, and headed for Antwerp to cut off our supply port and supply lines. German infantry and tanks attacked along a 90-mile front in an effort to split Allied armies on the Western Front. This caused a "bulge" in our lines, and the battle to flatten out this bulge by Allied troops for the next month was to become known as the Battle of the Bulge.

One of von Rundstedt's spearheads was headed for Luxembourg City with the objective of taking the city and the famous radio station known as Radio Luxembourg. The 80th Infantry Division was given the mission of holding Luxembourg City at all costs. I believe Patton said, 'Hold to the last man.'

My father's unit, the 80th Division, loaded onto trucks and within twenty-four hours, traveled 150 miles.

We rode in open vehicles. It was freezing cold and there was no way to keep warm. We had a few blankets and huddled close together to try to keep warm. There were no rest stops and no delays of any kind. When men had to relieve themselves, they did so off the back end of the truck, while trying to hang on at the same time.

We dug in near Junglinster, Luxembourg, just north of Luxembourg City, about eight miles from the front line. There, we found remnants of every outfit imaginable—a dozen from a rifle company, a few quartermaster trucks with their drivers, kitchen personnel, wiremen, artillerymen—all of them who had been separated from their outfits when the Germans launched their attack.

Over the next few days, the 80th moved every day to a different location (Steinsel—Beringen—Bourschied), traveling by truck and marching by foot, up and down steep, slippery hills.

The 80th attacked the Germans multiple times, pushing them back, but losing men from wounds as well as frostbite. At this point, the regimental battalions began to split up, each with a different mission.

Sleep was impossible because of the cold. I lay down beside a tank with Lt Bill Mounts and Kad (Lt Saul Kadison). It was about 3:00 a.m. It was still too cold to sleep. After a while, I tried to get up, but my coat was frozen to the ground. I had to wake Kad to help me get up. When daylight came on Christmas Eve Day, the cold was unbearable beyond description.

After a little while, a Kraut automatic rifleman opened up on the boys who were desperately trying to dig in between us and a nearby farm house. It was suicide for them to move. Some of them tried to come back to where we were behind a little knoll, but only a few of them reached us. Several were killed.

The Krauts started throwing mortars at us. That was rough, because we were close together in the open and had little cover. We couldn't see who was shooting at us, but we threw lead into the woods to try to keep the Krauts quiet while our boys out in the open worked their way back behind the knoll.

Our rations had run out the day before and we were getting hungry. It was cold and the wind kept blowing snow all over the place, making it difficult to see.

At first light, we got orders to saddle up and move back. We were to dig in on the high ground about a mile north of Niederfeulen, Luxembourg.

What a way to start Christmas Day.

The head of our column moved slowly and, as a result, the men began to bunch up. The Krauts noticed this opportunity and opened up on us with artillery as we moved up the side of a hill. They were firing "screaming meemies"—large-caliber rounds fired in clusters. We heard the rockets coming, but there was little we could do other than spread out and run like hell. There was no cover around.

Several men were hit. Some of the men hit the ground and wanted to stay there. That would have been suicide, but it was tough to get them up and keep them moving. When we got over the top of the hill and out of sight, the Krauts stopped shooting at us.

It was Christmas Day. Kad and I were determined to get our men warm. We went to investigate a nearby house and found that an antiaircraft outfit had been there before the breakthrough and, from the equipment and supplies left behind, it was easy to see that they had moved out in a hurry.

My feet were swollen and hurt like hell, but so did everybody else's, so I couldn't complain. We were to sit tight and wait for orders. We were also told that the kitchen was going to bring us up a real Christmas dinner with all the trimmings.

The men were tired and gloomy, their spirits were low, and their feet hurt. When I told them about the Christmas dinner, they all brightened up a little.

About 1:00 p.m., one of our cooks brought us our Christmas dinner. We had turkey and cranberry sauce with all the trimmings, and all you could eat, too. I thought about the 1st and 3rd Battalions. They had moved out to attack Kehmen at daylight. I hoped they might get some relief and warm food this Christmas Day.

My dad wrote the following letter to his parents the next day, Tuesday, December 26, 1944.

Dearest Mom & Dad,

Thought about you yesterday and was in hopes that Hazel and Lois (Dad's sisters) were able to spend Christmas with you. My Christmas was very enjoyable under the circumstances. My men and I spent our time indoors in a nice big Chateau where we had fires. We had turkey with all the trimmings for dinner. Our kitchen fixed it for us. My men got lots of packages and letters. I still haven't received any mail but eventually it will catch up with me.

Am enclosing a few clippings you can put them away for me.

The scenery here is beautiful. Very hilly and the ground covered with snow. I'm taking some pictures. The people are very friendly and do all they can do to help us. I'm feeling swell and am happy.

Devotedly,
Andy

Like so many men of "The Greatest Generation," my dad didn't want his parents to worry about him. You can certainly see the difference between what he (and many others) faced during that Christmastime in 1944 and what he wrote to his parents.

My dad was—*and still is*—my hero.

Never Forget the sacrifices our brave men and women the world over made (and continue to make) in order for us to enjoy our freedoms.

I wish you & your families a Merry Christmas and a wonderful holiday season. And as I write this from my Florida home, may your Christmas be warm, cozy, and full of love and good cheer.

(9) A Veteran's Christmas

First of all, Merry Christmas from Gainesville, Florida.

I have friends and extended "families" around the world. Some are veterans, some are classmates, some are business associates, but *all* are friends. I often call these wonderful folks, "my family away from home."

And like a few of my friends, if you don't celebrate Christmas, I wish you a warm, happy, and healthy Holiday season.

My mom kept all the letters I wrote home while in the Navy; just like my grandmother kept my dad's letters he wrote home during World War II. Recently, I read through some of my old letters, trying to find one that I wrote during the holidays I spent away from home.

During my four years in the Navy, I spent several Thanksgivings and Christmases away from home. When I flew to NAS Agana, Guam, I arrived the day before Thanksgiving. I was 18 years old and it was my first Thanksgiving away from home.

Number two was also in Guam. And both of my Navy Christmases away from home were spent in Guam.

I was an aircraft firefighter in what was called, "Crash & Rescue." At the Crash Barn, which was located adjacent to the airport runways, we worked 24-hours on and 24-hours off. Or, as the Navy conveniently puts it, "Port and Starboard." I was in the Port section. That was our working schedule for my 15 months in Guam.

I can't recall if I was on duty or not during the holidays. But chances are, I worked more than one.

For a lot of active duty personnel (then *and* now), Christmas was a little different holiday than Thanksgiving; I think so for almost everyone. People celebrate Christmas and Hanukkah in different ways. And some people don't celebrate at all.

Yes, *I know* they are two different holidays, but stay with me here.

What I mean is that for most of us growing up (that includes *YOU*), holidays were usually spent with our families. It didn't matter your race or your religion. The fact is, we all grew up with a family holiday tradition of some sort, year after year.

So, for many of us veterans who grew up with our own *individual* family traditions, our first Christmas (or Hanukkah) away from home was spent celebrating with complete strangers—all of whom had their own family holiday traditions.

At the time we may not have known it, but... we were family.

I'm going off on a slight tangent here and I know I'm stretching it a tad, but stay with me.

Imagine the board room at General Mills many moons ago, when someone brought up the idea of combining a *new snack*... with nuts... and pretzels... and while we're at it, let's add in some Worcestershire sauce, some butter, and don't forget, garlic seasoning.

Here's a bunch of different textures, flavors, colors, tastes, and aromas all thrown into a bag that will, over time, marinate. Or, as my mother used to say about her casseroles, they "have time to get acquainted."

That's what we were. Guys from all over the country with different tastes and colors, different religions and traditions, and yes, some of us had a stronger aroma than others. We were thrown together into a cohesive, working, unit—a *family* of sorts. And "we got acquainted."

As I think back on those first holidays away from home—in the military—one thing I do remember is missing the time together with my family at home. Most of the times, I remember lots of family, lots of good food, and lots of happiness.

Celebrating Christmas in Guam was, how shall I say, *different* than celebrating at home. If on duty at the Crash Barn, we shared a good meal with our new family. Several of us played guitar and we may have even broken out in a Christmas carol or two. The food was good and there was always plenty of it. I don't ever recall going hungry in the Navy.

But I know it wasn't that way for everyone, especially those in the middle of a war—yes, we were still fighting the Vietnam War.

One thing was certain, though, and I don't think it mattered if you were Army, Navy, Air Force, Marine, or Coast Guard. It didn't matter if you were in a foxhole, a hooch, or on a South Seas island. When you got a package of goodies in the mail during Christmas, *everyone* shared what they had. There's probably a standard issue General Order somewhere that states, "thou shalt share all holiday goodies with other members of your unit."

That was a military tradition, carried on for generations. My dad even mentioned this several times in his letters.

I learned a lot during my time in the Navy. One of the most important things I learned, was that while we all came from different backgrounds and traditions, we all served together, as a team, with the same basic principles and ideals.

And that, my friends, is one of the most honorable traditions I can think of.

During this time of the year, I know you'll remember that there are hundreds of thousands of service men and women who are not able to spend this holiday time with their own families. Some may have a tougher time celebrating than others. But they are out there, helping to protect the very freedoms we continue to enjoy today.

If you think about it, all of us are connected to a veteran in one way or another. Our grandfathers may have fought in WWI or WWII; our fathers may have fought in WWII or Korea or Vietnam. We certainly have classmates and friends who joined and served (or are currently serving). We owe these men and women a rousing round of thanks for what they do.

WE… ARE… FAMILY!

Merry Christmas, Happy Hanukkah, and Happy Holidays.

(10) You Light 'em, We Fight 'em

Picture this: the wide, crystal-clear, deep blue ocean… clean, crisp air… and beautiful sunrises and sunsets. All easily seen from the flight deck of an aircraft carrier with a 60-foot drop straight to the ocean surface.

Working on the flight deck of an aircraft carrier has often been called, "the most dangerous job in the world."

A mixture of beauty and danger, every second of the day—every 20-year-old's dream. Well… maybe not everyone.

Of course, back in 1975, it didn't seem that dangerous to me. I worked on the flight deck of *Kitty Hawk* in the Crash & Salvage Crew. It was just another day on the job. Or as we veterans often said, "Another day in paradise, livin' the dream!"

The job of Crash is simple: sit around and wait for an aircraft to crash and when it happens, try to put the fire out quickly and rescue the pilot and crew before things get really scary.

Our motto was, and continues to be, "You light 'em, we fight 'em."

Most of the time during the Westpac '75 cruise, flight ops ran from 10am until midnight. Crash had to be on deck an hour before and stay until an hour after.

At that time, there were 14 of us in Crash. Navy regs required a minimum of 11 Crash crew members to be on the flight deck during flight operations. It made for a long day, but it was our job and we did what was needed and most of the time (well… *some* of the time), we didn't complain.

It was not uncommon for us to pull double, rotating shifts during extended flight ops, which happened more often than you'd think.

Flight ops became a rewarding and challenging experience and sometimes, an "interesting" routine. Stay with me here.

I learned to drive Crash firetrucks when I was stationed at NAS Agana, Guam. That was fun back then on Guam, but as you can imagine, even more challenging on *Kitty Hawk*.

The main difference between driving a Crash firetruck in Guam (they were the same model, an Oshkosh MB-5, for those keeping track of such things) and driving one on an aircraft carrier was, well, tight spaces. NAS Agana, Guam had hundreds of acres of paved runways, taxiways, and side roads—a "luxurious" drive in the country.

Kitty Hawk flight deck was a little more than four acres with upwards of 50 aircraft parked and constantly moving. No matter where I drove or parked the Crash truck, I was always in the way.

As a teenager, growing up in Gainesville, Florida, I wasn't that careful of a driver. I had four accidents and three speeding tickets before graduating from high school. But who's counting?

Of course, *none* of them were my fault.

But something about driving a fire truck… on the flight deck… of an aircraft carrier. Yeah, I grew up quick and became a more responsible driver.

One day at sea during the Westpac '75 cruise, I was driving the Crash truck; it was during daytime flight ops. I was parked just aft of the "waist" launch catapults (amidships), because aircraft were being launched off the forward bow catapults.

An A-7 Corsair was hooked up on Cat #2, ready to be launched. Everything looked normal and the plane was armed to the gills with both missiles and bombs. The Jet Blast Deflector (JBD) was up, the plane at full power, and the Shooter (Catapult Officer) signaled to launch the plane. The launch was smooth except for one thing—the A-7 dropped one of its 700-pound bombs on the flight deck on the way out. It skidded to a halt about fifty feet short of the ship's bow.

Everybody in my truck saw it and there was one big collective, "Oh crap!" Actually, we said something else, but I won't repeat it here.

The V-2 Division catapult crews cleared the flight deck, heading in all directions. Fortunately, none jumped overboard. We didn't know how these bombs worked, whether this one was armed and ready to explode, or whether we should bend over and kiss our butts goodbye.

Bos'n Robuck (the Crash Officer) was the first to approach the bomb.

Let me pause here a moment and tell you a little about Bos'n Tyrone ("Ty") Robuck. He was one of the officers I admired the most while aboard *Kitty Hawk*. He was a Chief Warrant Officer (CWO-2), a proud Cajun, originally from Louisiana. Prior to the Navy, he was a former Marine, who (I believe) had two tours in Vietnam. He wasn't a big man at all, physically, but he had mustered up a chest full of ribbons. So many, in fact, that he walked with a port list (ribbons were worn on the sailor's left side). He was confident, he was fair, he was a good listener, and a great leader.

Bos'n Robuck casually strolled over to the bomb and looked it over. He then turned, looking straight at me, and signaled for me to drive the Crash truck up closer. I shook my head and politely mumbled, "Hell no! That's a live bomb." Actually, I said something else, but I won't repeat it here.

Within a few minutes, two EOD (Explosive Ordnance Disposal) team came onto the flight deck, examined the bomb, and declared it safe—it had not yet been armed.

What did we do? Another Crash crew member (Gary Borne, I

believe) drove the Crash forklift over to the bomb. The ordnance guys helped move it onto the forklift and Gary was directed to dump the bomb overboard. Everything was suddenly routine again and we got back to regular flight ops.

Total time for this little "challenge" was less than 10 minutes.

Just another fine day in the Navy.

Several points to make here.

The average age of men (and now women) on the flight deck is 19; I was 20 when I first arrived aboard *Kitty Hawk*. It is a time in my life that I will Never Forget (see how I cleverly slipped in the title of my new yet to be published book).

Where were you and what were you doing when you were 20 years old? Go ahead, I'll wait.

Even though we trained constantly for different situations—putting out aircraft fires and rescuing pilots and crew members—how do you train for a bomb dropping on the flight deck? You really can't. But your training, confidence, and experience automatically kick in and you react as best you can as a team.

I sometimes "joke" about the above story, but it really happened as written. My good friend and shipmate, Glenn Law, was the Yellow Shirt Director on Cat #2 that day. He later told me that he ran like hell toward the fantail when the bomb dropped. I think he said, "I made Bob Hayes look like a turtle." (Bob Hayes, by the way, was an Olympic sprinter and at one time, the "fastest man alive.")

Fortunately, on that occasion, there was no explosion or death and no one was hurt. That wasn't always the case on the flight deck. But that's another story to be shared at a later time.

Most veterans periodically think back on their time in the service. Some have good memories, others don't. But we all have those memories. We all served our country the best we could with what we had. I like to think that we're all proud of that service.

When I see another veteran (usually identified either by uniform, by a ball cap, or with a certain type of tattoo), I try to acknowledge their service. Sometimes they want to talk a bit, other times they are just appreciative.

When you see a veteran, if you thank them for their service, they're, of course, appreciative. If you really want to thank them, take a moment to ask them about it. And… listen.

(11) Does Anyone Write Letters Anymore?

Letter writing has become a pastime; almost obsolete for today's generation. But back in the day—that is, when I was in the Navy in the mid-70s—letters from home were always welcomed. And in the same vein, letters I wrote were basically my only communication with my family and friends.

I'm willing to bet that most people still like receiving letters & cards today. I know I do. But our "instant" society typically uses short emails and even shorter texts to communicate. That's not necessarily a bad thing, but it does take away from the real heart of writing *and* receiving letters, especially to an active duty member of our military.

When was the last time *you* wrote a letter… a *real* letter?

Dear Son…

During my short, 4-year stint in the Navy, my mom wrote a letter to me almost every week. My dad was a lawyer and always had plenty of legal-size yellow pads—that's what my mom used to scribe her weekly "newsletter."

She'd usually start writing the letter on a Monday or Tuesday and write a little bit almost every day, filling me in on what was happening in their lives, my sister (who also wrote regularly), and friends and other family members.

Then, she'd finish it up on the weekend and get it in the mail first of the week. Sometimes, my dad or another visiting family member would add in a paragraph or two—that was special. Many of these letters, all handwritten in my mom's beautiful cursive script, would top 10 or more pages. It was almost like a long conversation with my mom.

Those cherished words, "Mail Call!"

Depending on where I was at the time (NAS Agana, Guam, or at sea aboard *Kitty Hawk*), the mail would take anywhere from a week to several weeks to reach me. When you're on a ship in the middle of the Pacific Ocean, the Navy flies a plane out to deliver supplies, which included mail. If we were near a land airbase, there would be a daily flight. Otherwise, it may be a few days or a week before receiving mail.

Once the mail was delivered to the ship by the COD (Carrier on-board Delivery) aircraft—usually in large canvas bags—it would be taken down to the ship's post office and sorted by division. Someone from each division would head down to pick up the mail, bring it back to the division office, then further sort it before distributing. Keep in mind, there were over 5,000 sailors aboard *Kitty Hawk* during the cruise.

When you're on active duty—it doesn't matter if you're Army, Navy, Air Force, Marines, or Coast Guard—there's *nothing* like hearing, "Mail Call!" and getting a letter from home. No matter how crappy a day you may have had, letters from home were *always* the highlight of the day.

The only exception I can recall was a "Dear John" letter. I know a few guys who got one and it was usually obvious by their change of demeanor. But they eventually got over it, especially when we hit the next port call.

I still have the letters I wrote home (*45+ years ago*). My mom kept them for me. I recently re-read the first letter I wrote from boot camp, just after I arrived at Orlando RTC.

I'll spare you the details, but the letter begins, "This place around here is a hell hole. We have to ask for everything, or be given an order to do anything."

Yep, that pretty much summed up boot camp. But when I got out of boot camp, out of "A" school (Lakehurst, NJ), and into the "real" Navy, things dramatically improved. I wrote about my boot camp experience earlier.

The last few letters I wrote home from *Kitty Hawk* showed a *slight* improvement. Well, actually, a grand improvement, especially since I had been promoted to a Second Class Petty Officer (ABH2). Of course, it helped knowing that I was coming home soon (30 days and a wake up; 20 days and a wake up; and so on).

Letters from WWII

Believe it or not, I also have my father's letters he wrote home to *his* parents during World War II (75+ years ago). He was an Army officer and didn't complain as much as I, an enlisted man, did. But still, reading his letters and knowing where he was and what he was doing when he wrote them, told me a lot about my father that I did not know while growing up.

There are a couple of passages in my WWII book where I my dad described a horrific battle, then I added in an excerpt from a letter he wrote home soon after that battle. Which, by the way, didn't describe *any* of the carnage he had witnessed. Back then, that generation simply didn't want their parents to worry during the war.

Unfortunately, I don't have those long letters my mom wrote to me. You bet your sweep bippie I wish I did. But there was limited storage room on the ship (we slept in racks, stacked three-high—mine was the middle rack, just because I know you wanted to know). The only storage we had was limited to our rack area and let me tell you, there wasn't any extra room. So, I had to toss out my mom's letters.

Now, email rules…

Today's Navy—actually, *all* the military—now has the ability to receive almost instantaneous communications from home. Whether it be by cell phone, email, or text, our military personnel can have that personal connection, only a little quicker. I still like letters, cards, and packages, but I can certainly understand the differences available today.

Back in my day, if you were a sailor out on the big blue and your wife was expecting a baby, it would be a week or so before you'd hear (unless she or your parents sent a telegram). But it would still be a while before you saw the first pictures of your brand-new son or daughter… or twins! Now, you can get email and text with photos of your newborn!

To counter today's "instant communication," let me share with you another story. Many of the *older* veterans will most certainly relate.

My first duty station was NAS Agana, Guam; I was there for 15 months. BTW, that's an island in the Pacific Ocean out in the middle of *nowhere.*

Back then, several college age women friends wrote to me (they were still attending FSU). Something about writing to a sailor seemed "romantic" at that time… But I digress.

My female friends may have dabbed a little bit of perfume on their letters. That was *always* welcomed and of course, the other guys in the division would want to know a little more about that particular young lady.

I say that, because until they invent a "scratch & sniff" email delivery system, *nothing* will replace getting letters from home.

To answer the question at the beginning, "Yes, they do."

I work closely with the 80th Division Veterans Association. Besides creating the website (www.80thdivision.com) and digitizing more than 100,000 pages of reports & photos to place online, I am also one of the association historians—I focus on WWII.

A month or so before Veteran's Day 2019, Jason Shelton, an 8th-grade history teacher in Greenville, TN, contacted me with an unusual request. His students were going to write letters to WWII veterans as part of a history project. I provided him with the names of those with the 80th Division. A few weeks later, Jason let me know that his students wrote and mailed a total of 334 letters to World War II veterans.

One more… my beautiful and gracious 98-year-old mother-in-law has written letters all her life. Over the past few years, her letter writing has become less frequent, primarily because of, well, getting older. But she still insists on sending *Thank You* notes to friends and family who visit and send her cards.

While she needs help addressing the envelopes and her hand-written notes are not nearly as long nor as neatly written as in years past, the simple fact that she continues to see the importance of letter writing leaves me, in a word, speechless.

So, the answer to the question at the beginning of this blog is a resounding, "Yes!" People still write letters. Awesome!

(12) Writing Your Own Memoir, Part 1

It's Really Easy, and… What a Legacy

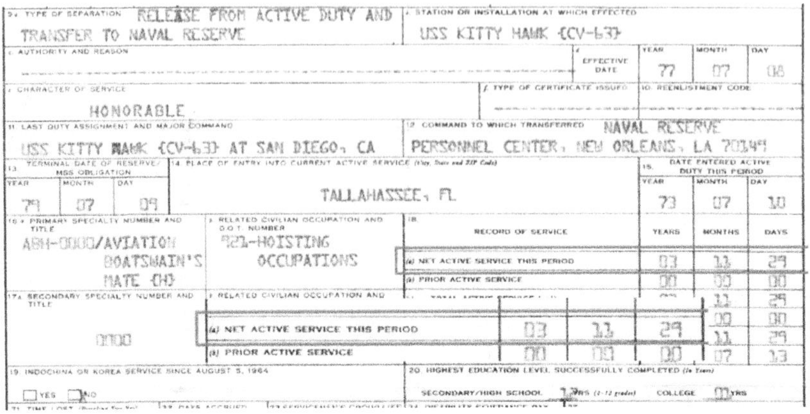

9a. TYPE OF SEPARATION: RELEASE FROM ACTIVE DUTY AND TRANSFER TO NAVAL RESERVE
b. STATION OR INSTALLATION AT WHICH EFFECTED: USS KITTY HAWK {CV-63}
c. AUTHORITY AND REASON
d. EFFECTIVE DATE: YEAR 77 MONTH 07 DAY 08
e. CHARACTER OF SERVICE: HONORABLE
f. TYPE OF CERTIFICATE ISSUED
10. REENLISTMENT CODE
11. LAST DUTY ASSIGNMENT AND MAJOR COMMAND: USS KITTY HAWK {CV-63} AT SAN DIEGO, CA
12. COMMAND TO WHICH TRANSFERRED: NAVAL RESERVE PERSONNEL CENTER, NEW ORLEANS, LA 70149
13. TERMINAL DATE OF RESERVE/MSS OBLIGATION: YEAR 79 MONTH 07 DAY 09
14. PLACE OF ENTRY INTO CURRENT ACTIVE SERVICE (City, State and ZIP Code): TALLAHASSEE, FL
15. DATE ENTERED ACTIVE DUTY THIS PERIOD: YEAR 73 MONTH 07 DAY 10
16a. PRIMARY SPECIALTY NUMBER AND TITLE: ABH-0000/AVIATION BOATSWAIN'S MATE {H}
b. RELATED CIVILIAN OCCUPATION AND D.O.T. NUMBER: 921-HOISTING OCCUPATIONS
17a. SECONDARY SPECIALTY NUMBER AND TITLE: 0000
b. RELATED CIVILIAN OCCUPATION AND

18. RECORD OF SERVICE	YEARS	MONTHS	DAYS
(a) NET ACTIVE SERVICE THIS PERIOD	03	11	29
(b) PRIOR ACTIVE SERVICE	00	00	00

19. INDOCHINA OR KOREA SERVICE SINCE AUGUST 5, 1964: YES / NO
20. HIGHEST EDUCATION LEVEL SUCCESSFULLY COMPLETED (In Years): SECONDARY/HIGH SCHOOL 12 YRS (1-12 grades) COLLEGE 00 YRS

Note: This is the first of a three-part series article. Next week, ***Research: After 30+ years, I'm back on the flight deck.***

Several years ago, I wrote and self-published a book about my 4 years in the Navy. It initially started out as a simple memoir. But over time, it morphed into a full-size book. I did not write the book to make money. It was not my intention to sell it, but to *preserve.* My sole purpose was to share my Navy experience with family, friends, and shipmates.

Yes, it is on Amazon. No, I didn't sell a lot of copies. It's a friggin' *memoir*, not a romance novel. I wanted to tell it like it was (for me), "sprinkle" in a few facts, and elaborate on a few stories. There are probably still a few spelling and grammar errors (I didn't use a professional editor), but I'm okay with that.

When I received the first "proof" hard copy in the mail, I was impressed at just how easy it was. I will tell you this, though. It certainly helped me along in my current writing "journey."

It's a "process," a true labor of love

This blog article is NOT about *my* book; it's about the *process* of writing it—something *you* may want to consider. Think about it… a legacy for your family, your kids, and your grand kids. Best of all, it's not that hard. And while you read through this blog, you may also realize you can help write your father's, mother's or grandparent's memoir. It doesn't have to be about you!

In 1973, I signed up for a four-year stint with the Navy. But because I went in on a Monday and got out on a Friday, I *almost* made the complete four years. The title of *my* book is, ***Three Years, Eleven Months, 29 Days—But Who's Counting***. I just couldn't resist that particular title. And if you question my "exact" time in service, I've provided a snippet of my DD-214 discharge papers as proof.

Writing my memoir wasn't difficult at all. Yes, it took some time and I drifted in and out of writing. All in all, it probably took me about a year to write, from the initial outline to the first proof copy. But it was certainly worthwhile.

You don't have to have a literary agent or a publisher to write & publish your memoir. I know a lot of men & women, WWII, Vietnam-era, and Global War on Terror veterans, who wrote their memoir either using paper and pen or a computer and only had a few copies printed for their families and friends. For each, it was a most fulfilling experience and I don't know any that regretted writing it.

There are methods, however, to self-publish, and I'll share some of those with you later in Part 3.

I self-published my book using Amazon's CreateSpace (now known as Kindle Direct Publishing), but there are other self-publishing services. It only cost a few bucks—for the proof copies and for extra copies I ordered for family and friends. More about that later.

The "Why" of writing your memoir

Serving in the United States Navy was one of the most honorable and memorable periods in my life. I find that as I get older, I cherish those memories even more. I am proud to be a veteran. I also have a much higher respect for our men and women in uniform who currently serve than I did back in my day. I don't know if that's age or wisdom, but at this point, it really doesn't matter.

I realize others may *not* feel the same way about their service, and that's okay. Of course, there were times when I'd complain. But for me, I had more good times than bad, and that's what I always try to focus on.

A lot of what I learned in the Navy continues with me today, including some of the military lingo. I'm sure there are veterans out there who also continue to use some of those "choice" terms and language. Face it, it's a culture and tradition we want to continue.

I was discharged in 1977. But I didn't write this book until 2014, though I had thought about writing it for years. But you know… life got in the way.

When I did start to actually write it, I still worked full-time. But I managed to find the time (nights and weekends) to crank it out. Our kids had grown, moved out, and finished college, so I had a little more "Andy time." And the process of "journaling" my service (i.e., writing my book), both good and bad, was for me… both enlightening *and* memorable.

I know some people, including my own father, who felt "closure" after completing their book. I'm not sure I felt "closure"—I wasn't looking for it. Even though the Vietnam War was winding down when I served, I wasn't in a war, like my dad. But I know other veteran authors who were and for them, writing their book *did* bring closure to a chapter in their lives.

Writing Your Memoir

I think most memoirs are a "from here to there" type of book—events that happen in chronological order. That's how I wanted to write mine, plain and simple, from start to finish. It also provided me with a way to "relive" my Navy days. That was fun!

I wanted to write about boot camp, "A" school, and my two duty stations, NAS Agana, Guam and *Kitty Hawk*. Those topics became my "sections." And of course, I wanted to write about the Westpac '75 cruise.

It's hard to explain to someone what it's like to be aboard a Navy ship during a 6- to 8-month cruise. In my case, I worked on the flight deck of an aircraft carrier, *USS Kitty Hawk*. But it was important to me to describe the comradery and friendships that develop when you're together for that period of time. In my case, it was also the most fun part to reminisce and write about.

I also wanted to include enough details so that someone who had *not* been in the Navy could understand the culture, as well as appreciate what we did on a day-to-day basis. Most of this I knew from experience, but in some cases, I needed to do additional research. Lo and behold, Dr. Google and Mr. Internet.

Because I had written a military book several years earlier—about my father's World War II experience in Europe—I had an idea of how to tell the story and where to find research materials. For *that* book, I had also worked with both an editor and a publisher.

But back then (2005), the self-publishing industry was just ramping up and still had a ways to go. For my WWII book, I found a traditional publisher, Casemate Publishers, who is still going strong.

For my Navy book, I *knew* I would self-publish—a totally different world and when the time came in 2014, most of the self-publishing industry quirks had been worked out.

Research—My Old Letters and My Photo Albums

A lot of my "research" came from my old photo albums and the letters I had written home—my mother kept all my letters. Those were treasures and provided me with not only the timing of certain events, but also how my attitude changed as I "matured" into an experienced and seasoned sailor.

Boy, some of those letters I wrote back then really told it like it was. But I digress…

I know a lot of veterans who have the letters they wrote home to their parents, their wives, their children, and other family members. That's a great place to start.

Outlining Your Memoir

The first thing I did was to outline my time in the service. As I mentioned above, my main "sections" were:

- Joining the Navy (the "decision" to join and the Induction

Center "experience").

- Boot camp.
- "A" School (Aviation Boatswain's Mate—Handler; Lakehurst, NJ).
- NAS Agana, Guam; both on-duty activities as well as off-duty.
- *USS Kitty Hawk*. I thought it would be useful to not only describe what I did on the ship, but also provide a "tour" of the ship and the flight deck and describe flight operations in detail.
- Westpac '75 cruise; what we did during the cruise as well as the ports we visited. Well, at least what I remembered, if you know what I mean.
- Discharge.

That was pretty much it. Once I had this initial outline, I just started writing, knowing that I would need to "fine-tune" the grammar & spelling at a later time. My wife and my sister also helped me with the editing.

The secret? Just start writing and know that you're not going to finish this overnight, over a weekend, or within a month. It depends on whether you want to write about your complete Navy experience from start to finish, or if you want to only write about a few things: your choice. Both will be treasures for your family.

I didn't write from beginning to end. I decided to first focus on certain times or stories, then expanded from there.

Warning: Navy "Language" ahead

As I began to write, I found my "voice," so to speak. You may not know what that means now, but you will once you start writing. For me, that simply meant that I wanted to not only tell it like it was, but to insert certain military and Navy terms along the way, with explanation and definition.

After a while, I realized that I needed to put a disclaimer on the book. So to add in a little humor, I added in the tagline: "Warning: this book contains three languages: English, Navy, and Military.

Some readers may find two of the languages offensive." I even included a glossary of terms, something I thought would be helpful.

Writing Prompts

I know several of you have probably *thought* about writing… something… about your time in the service, or perhaps your parents or your grandparents. But in the same vein, you may also be asking yourself, "How do I begin?"

Here are a few "writing prompts" that may help you get the ball rolling. Simply take one and start writing. It really is that simple. And what's best? You'll not only surprise yourself at what you write, but you may also be well on your way to writing your memoir.

While these writing prompts are geared toward the military, you should be able to adapt them to almost any style memoir. I'll provide three prompts for each of the blogs in this 3-part series:

I joined the Navy (Army, Marines, Air Force, or Coast Guard) because (*why?*).

I went to boot camp (*where?*). The thing I remember most about boot camp was (*what?*).

When I arrived at my first duty station, I thought to myself (*what?*).

Next week, I'll detail some of *my* additional "hands-on" research. I asked the Navy if I could visit and observe boot camp; I asked if I could spend time on an active carrier during flight ops. Did they allow me to do those things? Yes, they did! And more…

(13) After 30+ Years, Back on the Flight Deck, Part 2

Last week, I blogged about "why" you may want to write your own memoir. I also provided some simple tips, outlines, and writing prompts to help you get started.

While I'm a Navy veteran, my dad was Army (80th Infantry Division) as well as my father-in-law (82nd Airborne Division); both fought in World War II. We had many *interesting* conversations before they passed away. I wrote a book and an oral history about them. That helped me better understand how they dealt with their own war memories.

I know that my own children *and* their children will now know more about their fathers and grandfathers because of these priceless memoirs.

I encourage you to do the same—write your own memoir. If not for you, then for *your* children and *their* children.

You don't have to be a veteran to write or capture a memoir. What about your parents or grandparents? Even though they may not have served in the military, they've still experienced life—in some cases, a long life—and they've got a lot to share… in their own words.

Several years ago, my wife conducted an oral history of her mother and learned a few things she didn't know growing up.

My gracious mother-in-law will be 99 years old next month and while her body is not as strong as in past years, her mind is still sharp. When she and her *younger* 90-year-old sister get together, it is a blast to watch and listen to them reminisce about growing up during the depression in Boston. So it's important to capture those moments *and* that history; you may not get another chance.

Stay with me here. Back to my memoir…

All I had to do was ask

Soon after I started working on my memoir, I had a *brilliant* idea. Since I was in the Navy in the mid-70s, I thought it would be interesting to observe and compare today's recruits and sailors, both during boot camp as well as onboard a ship.

All I had to do was ask.

At the time, there was a Navy Public Affairs office located in Manhattan. I happen to be speaking at a legal technology conference in New York, so I made an appointment to meet with one of the Public Affairs officers. I had previously written a letter to them. But what the hell, a face-to-face meeting certainly couldn't hurt.

I asked for three things:

To spend time in boot camp, to observe new recruit training, inspections, and graduation.

To spend time aboard an active aircraft carrier (preferably out of San Diego), to observe flight ops.

To land on a carrier and to launch from the carrier.

Guess what? The Navy granted all three wishes. There truly is a Navy genie… But I digress.

Boot Camp, Then and Now

I spent three days at Great Lakes RTC and was able to not only observe, but also interview both Recruit Company Commanders and a few recruits. I wanted to compare and contrast today's recruit training to what I went through in the early 70s.

I was able to observe new recruits arriving off the bus (at midnight); receive their sea bags and uniforms; observe inspections and training exercises; and best of all, attend graduation. I'm a sucker for military traditions and watching these young new recruits graduate was a great experience.

A few things were obviously different. I went through Orlando RTC in '73. At that time, there were three Recruit Training Commands: Great Lakes, Orlando, and San Diego. Now, there is only one: Great Lakes RTC.

My "research" took me to Great Lakes, Illinois in 2009. In the 70s, we had one Company Commander for 80 recruits. When I interviewed CCs at Great Lakes in 2009, there were three Company Commanders assigned to a company of 80 recruits.

Remember the "deck parties" during boot camp? You know, pushups, sit-ups, and those hellacious, "Hello Dollies," which seemed to last for hours. Aching muscles and sore backs—built character, didn't it?

Well, today's CCs have an "exercise card" they carry with them, indicating the "limit" of daily recruit exercise time. I don't know exactly what those limits are, nor do I really care. But the fact that there are now *limits* to the amount of PT per person per day is…. Well, I simply won't go there.

Also, while we had female Navy recruits at Orlando RTC, they were berthed and trained in a different part of the Center. The only time we'd see them was during church services, which made for some *interesting* "sermons."

Now, male and female recruits train together. IMO, that's a good thing, especially because they now serve together onboard ships.

The PAO (Public Affairs Officer) was gracious and accommodating and I was very appreciative of the time she provided to me, as well as being able to observe almost all activities, especially the graduation ceremonies. I love military traditions and was not disappointed.

Flight Deck Heaven

For those of you who served in the Navy, how many have yearned to go back aboard a ship… at sea… one more time? I know there are many, especially us older veterans. The younger ones? Well, maybe not so much. But eventually you will. I can almost guarantee it.

When I was discharged in July '77, all I wanted to do was to put as much distance between me and *Kitty Hawk* as I could. I was "done" with her and I was "done" with the Navy. But for the past 25+ years (before writing my memoir), all I wanted to do was to spend one more day with her.

Anyone else feel that way?

In March 2009, the Navy flew me out on a C-2 Greyhound COD (Carrier On-board Delivery), along with seven other VIPs (why they called us VIPs is beyond me), from NAS North Island (San Diego) to *USS Ronald Reagan*, CV-76. I spent two days aboard.

Let me tell you about it.

First of all, I figured they would let me observe flight deck operations from Vulture's Row, an outdoor observation balcony on the island several decks above the flight deck. Nope. My "guide" was a fellow ABH who was the Flight Deck Officer. He gave me a float coat and a cranial helmet and said, "Follow me!"

I was in AB "Hog Heaven."

After 30+ years, I was out on the flight deck again, right next to launching aircraft and right next to the foul line, watching recovering aircraft, just like the "old days." The sights, the sounds, and the smells took me back in time. The cool, clear breeze coming over the bow… just as I remembered.

And yes, I did walk up to the bow, leaned over the safety nets, opened up my jacket and "flew." Just like the old days. What a feeling!

Best of all? I had my camera. There were several areas I was told *not* to photograph, but none of those were of interest to me. I was *focused* (pun intended) on flight operations.

The major difference I noticed was that most of the aircraft were various models of the F-18 Hornets. I guess that makes sense for maintenance, parts, mechanic training, pilot training, etc.

But I missed the variety of aircraft from my day: A-7 Corsairs, F-4 Phantoms, F-14 Tomcats, A-6 Intruders, EA-6 Prowlers, S-3 Vikings, A-3 Skywarriors, and A-5 Vigilantes. They were all great aircraft and served a purpose at one time. My favorite back then? The A-7 Corsair.

Since I had been in V-1 Division/Crash & Salvage, another major difference I noticed was the Crash equipment. No longer MB-5s (I wrote about them earlier). Now, the Crash Crew rides a smaller unit, the P-25 Mobile Firefighting Unit. It's smaller and more maneuverable. But it's still Crash and they still the have same motto, "You Light 'em, We Fight 'em."

One more observation. Back in the mid-70s, there were 27 *active* aircraft carriers in service around the world; we had 14 crew members in Crash. We had to have 11 Crash crew on deck at all times during flight ops. That made for some long hours and at times, double shifts.

In 2009, aboard *USS Ronald Reagan*, I counted 26 crew members in Crash. I also note that the US Navy now only has 11 *active* aircraft carriers. So, more sailors are available for billets? I don't know the numbers, but it seemed more than a coincidence.

Getting There & Back

During my 2½ years aboard *Kitty Hawk*, I witnessed (up-close) literally thousands of launches and recoveries. I always wondered what it would be like to land on a carrier. And better? To experience a catapult launch.

Well, my wish came true. The landing was cool, but the launch… spectacular!

It wasn't a fighter aircraft or a bomber—it was just the COD. But to be able to experience a cat launch was something I'll *never forget.*

To this day, I can still "feel" that cat launch (in the COD, passengers face aft). But those two seconds of straining against the seat belts and shoulder harness toward the rear of the aircraft, holding your breath, sucking in your chest, and not wanting it to end… and then that "release"… well, those two seconds still stay with me, even today.

One more thing…

I figured since the Navy granted me *these* three wishes for writing my book (Boot Camp, Active Carrier, Land/Launch), I might look at writing another Navy book—this one about the Blue Angels [<wide grin>].

Writing Prompts

I mentioned in last week's blog about "writing prompts." The idea is that you have a story to tell, whether it's just a simple, short one-page story about a certain event, a one-chapter story about a tour of duty, or a start-to-finish memoir.

Here's a few more writing prompts for this week:

The Navy (or Army, USMC, USAF, or USCG) gave me the ability to (*do what?*).

My job in the Navy (or Army, USMC, USAF, USCG) was (*what? Provide the details and what you did on a daily basis*).

The most influential veteran I served with in my military service was (*who?*).

Since I've covered the basics of writing and researching your memoir, next week I'll cover the final step: ***Publishing your memoir.*** In this case, what does it take to self-publish?

(14) Publishing Your Memoir, Part 3

Over the past two weeks, I blogged about the (1) reasons you may want to write your own memoir and how I (2) asked the Navy to help me with my personal "research." I also provided some simple tips and outlines to help you get started, as well as several "writing prompts."

BTW, I've already received several "thank you, this is perfect timing" responses to these posts. I know many of you have thought about writing your own memoir, so hopefully, these posts will also help *you* get started.

In this week's blog post—the final in this 3-part series—I'll provide several ideas on how to self-publish your memoir, should you choose to do so.

Step 1: Write

First, you have to write something… *anything* (that's a given). I cannot stress the importance of passing along your legacy, your memories, and your stories to family and friends. Even if it's just a short story or two. And as I mentioned in the past, you don't have to be a veteran to write a memoir. Think about your parents or your grandparents. They all have wonderful stories, too, both growing up as well as experiences in their own lives. And if you have kids that want to help out, it's also a great way for them to bond with their grandparents.

Second, realize that once you start writing, you may want to end it after the first story/chapter/section. That's okay. You can always pick it up later. No one is forcing you to write this, so you're your own boss.

Lord knows how many times I said to myself, "*Self*... Screw it! (or some other *choice* Navy term) I can't do this right now." But I'd pick it up a little while later and continue writing. Mine started out as a simple memoir, but morphed into a whole book, taking about a year, from start to finish. I didn't know that at the time when I started.

Third, once you've written something, what's your next step? *That* is the real question that only *you* can answer.

Examples of memoirs

Let me share with you some things I've learned over the past 20 years of working closely with veterans and their personal histories.

My dad (80th Infantry Division) kept a diary during the war—a day-to-day journal of his experiences during World War II. He started writing while he was laid up in a hospital for two months in England recovering from a wound. I wrote about this and my own WWII "journey" earlier.

After my dad passed away in 1989, I started working on his diary and eventually turned it into a book: ***You Can't Get Much Closer Than This: Combat with Company H, 317th Infantry Regiment, 80th Division***. It was published in October 2005 by Casemate Publishers and the following month, November 2005, it was selected as the *Book of the Month* for the Military Book Club.

It's now in its second printing. I *never* thought...

My father-in-law, Dr. Rufus K. Broadaway, was an officer in the 82nd Airborne Division and dropped into Normandy early in the morning on June 6, 1944. He was involved in the bloody battle for the La Fiére Causeway, just a short distance from Ste. Mere-Eglise. And for a while, he was the *aide-de-camp* to General James Gavin, commander of the 82nd Airborne.

Rufus wrote several stories, not only about his airborne days, but also about his days as a surgeon in Miami. I also had an opportunity to conduct an extensive oral history of his time before and during the war.

In my work with the 80th Infantry Division, I have read dozens of personal accounts by WWII veterans, many of which are available on the 80th Division website. Several memoirs eventually turned into books (www.80thdivision.com/books.html), most of which were written by the individual veterans' kids (like me).

Journaling to help heal

Memoirs do not need to be published works. Most of the time, they are memories that are personal and kept within the family. I also know many Post WWII, Vietnam, & GWOT veterans who write primarily to "get it out of their minds," to "put pen to paper," or to "let go of the past."

Whatever the reason, writing things down (often called "journaling") can help you clear your mind. That's what many psychologists recommend for their patients, too.

It helps. It works. Journaling is *not* for everyone, but it *is* for anyone.

My point it this: if you have several pages, you can simply use your computer to print out copies, or you can take it to a copy/print shop (e.g., Kinko's or Office Max). You may only want to share with family and friends. You may want to just send an email. Whatever works for you, it will be the right thing to do.

On the other hand, if you have a hundred pages or more (or less), maybe you *might* want to self-publish it using one of the many online "print-on-demand" services. That's what I did. And that's what I want to share with you now.

Self-publishing Services

There are literally dozens of self-publishing platforms and services online; more than when I self-published a few years ago in 2014. Just Google "self-publishing" to find a list. While there are many, the following are what I would consider the top five.

My disclaimers are: a) I Googled these myself, but I was already aware of these and b) I used CreateSpace to publish my Navy book and the six 80th Infantry Division books (written by Robert Murrell, past historian of the 80th Division). I recently ported them over to Kindle Direct Publishing.

- Kindle Direct Publishing (Amazon)
- iBooks (Apple)
- IngramSpark
- Smashwords
- Bookbaby

It might be advisable to spend a little time researching these various services. While Amazon KDP holds the market share, the other self-publishing services may have more benefits that are closer to your own needs.

Book Publishing Templates

The two most common book template sizes are 8½" x 11" and 6" x 9". I've used both and let me tell you why.

For my own Navy book, I chose the 6" x 9" because IMO, that's the most common size paperback book. Most of the books I get from the library (both hard cover and paperback) are this size. It really works well for self-publishing.

For Bob Murrell's 80th Infantry Division series of books, I chose the 8½" x 11" size because, well, that's what Bob used. For many (*many*) years, Bob had his books copied and bound (plastic spine) to ship to customers. But copy prices went up and profits went down.

A couple of years before Bob passed away, he allowed me to digitize his books and put them on Amazon KDP to "print on demand." The primary reason was to keep the legacy and the memories alive, and to *not* lose this valuable history. All proceeds for Bob's books continue to go to the 80th Division Veterans Association.

Amazon CreateSpace/KDP (and the other self-publishing services) all have ready-made templates (Microsoft Word, PDF, etc.) in different sizes (e.g., 6"x9", 8½"x11") for you to simply download and use for your memoir. Whichever template size you choose, I don't think you can go wrong.

Voila! You're almost there.

You can create your own book cover, too. I used my own photos (or the public domain Navy PR photos) for my Navy book. Bob Murrell had already created the "logos" for the covers of his 80th Division books, so I simply used those that already existed.

There are tools to not only help you with the formatting & cover creation, but also to automatically review the uploaded book. There is a "preview tool" that lets you "see" how your book will print. These tools exist for color or black & white, for hard cover or paperback, and for electronic book readers.

And if you want, name your price for your memoir.

A quick word about ISBNs

Keep in mind when you use an application like Amazon KDP, they will require you to have an ISBN number; it's free and can be assigned automatically, but the ISBN number is assigned to the "publisher." That might be you (it was for *my* Navy book), or it could be for a service provider/consultant. That's fine, nothing wrong with it. But it may limit you in the future should you want to publish it with a traditional publisher. If you want, you can purchase your own ISBN number.

Without going into details about ISBNs, here's a link to an excellent short tutorial from Self-Publishing School about ISBNs and how to order your own. https://self-publishingschool.com/isbn/

It really is that easy!

I'll admit, I'm a computer guy and pretty comfortable with these types of tools. But if you need help, there are several self-help "tutorials" as well as a vast online community forum. And if you want someone to help with along the way, there are also consultants, both directly with Amazon as well as independent companies.

What about eBook formats?

If you want to make your book available electronically, there are several different formats. Of course, you can leave it in MS Word format or convert it to PDF format—that's fairly easy and I know a lot of people who do that. If you want to convert to an eReader format, Amazon KDP has an online tool for that.

I've also used an application called *Calibre*, which I find very helpful. Using *Calibre*, I can convert my book into a number of different eBook formats. This is helpful when I want to share my book(s) with people that use a Kindle or Nook (or other eReaders).

What about the costs?

If you self-publish and do the work yourself (write, edit, and upload to a self-publishing service), the cost is zero-nilch-null-zip; at least, it is on Amazon KDP.

The *only* costs for my Navy book were for the printed copies I ordered, both for the proofs and for the extra copies for family and friends. The cost was about $5 each (my book is 246 pages, black & white print, color covers). If you print your entire book in color, the costs will understandably be a bit higher.

And to answer your next question, I do have photos in my Navy book, but I chose to print them in black & white to save money.

Yeah… I'll use Amazon KDP again.

Writing Your Own Memoir—Step-by-Step

So with all that behind us, how do you write and publish *your* memoir? Here is Andy's *simple* step-by-step instruction:

Organize your memoir into "sections." For example (Navy), Boot Camp, "A" school, duty station(s). This is often called "outlining" your book.

Gather your letters, photos albums, and other memorabilia to help remember people, places, duty stations, and dates.

If you want to elaborate, provide further details for each section. For example, I wrote about what I did while on duty as well as what I did while off-duty. Well, at least those things that I felt I could publicly share [*(wide grin)*].

Write, edit, re-write, and then edit again. Give it some time—memories often trigger more memories. And you may find yourself wanting to include more.

Have someone else read what you wrote for two purposes: one, readability (this is how the story flows); and two, accuracy (you want to make sure the facts are correct).

Then, find a way to publish or self-publish. There are several self-publishing avenues. I used Amazon CreateSpace, now known as Kindle Direct Publishing.

Order some copies and pass them along to your family, friends, and shipmates. You don't know how much they will appreciate this gesture.

If I can inspire one or two of you to memorialize your "journey," then my task has been successful.

Writing Prompts

The past couple of blog posts, I've provided several writing prompts—the purpose is to help you get started in writing your own memoir. When you start working on your memoir, you may feel overwhelmed. Some of you spent an entire lifetime career in the military; others may have only spent two years. But you've still got stories to share. Here are a few more prompts:

When I left the service, walking off the ship / out the main gate for the last time, I felt like (*what?*).

When I got home, knowing I would not have to go back, I wanted to (*what?*).

When I think back on my time in the service (how many years ago), the thing that stands out most in my mind is (*what?*).

When you're done, or if you've already completed your memoir, drop me a line. I'd love to know who else had written a book about their service.

(15) Tattoo—Yes… or No

I have a tattoo and I'm damn proud of it.

No, the "Mom" tattoo is *not* mine; it's just a lead off image to this blog post. Though, "MOM" was pretty popular back then… that, and a Navy anchor on the forearm, like Popeye.

But I digress…

Back in "my day," (mid-70s) *not* many people outside the military had tattoos. Or, at least as I remember.

Now… there are TV shows (more than one) about "ink." I don't watch them.

But I have to admit that I am fascinated by the different types of designs and colors of today's tattoos. Some of these tattoo artists are, IMO, quite talented.

I'm not here to judge whether having a tattoo is taboo or not, whether it is fashionable or not, or whether you should or shouldn't get a tattoo. That's *not* my job. That's *your* own personal decision.

All I'm doing today is telling you *my* story—my "journey"—of *my* tattoo.

Westpac '75—Hong Kong

Even though I'd already been in the Navy for two years, I didn't get a tattoo until August 1975 when *Kitty Hawk* was on a Westpac cruise and ported in Hong Kong.

We were in Hong Kong for a total of six days. Hong Kong didn't have a pier big enough to dock the *Kitty Hawk*, so we anchored out in the harbor near Green Island and took ferries from the ship to the port.

The trip in wasn't too bad. But that trip on the way back to the ship… on a dark night… in rough water… rocking back & forth, side to side… after a night of heavy drinking… Let's just say it wasn't a pretty sight, and leave it at that.

But I digress.

We also had what we called "max liberty," meaning that you didn't have to come back to the ship unless you had duty. Lucky me, I pulled duty the first day in port, so I had to wait my turn to go ashore, so to speak.

But during those six days, I actually did some touristing while in Hong Kong (as opposed to parking my butt on a bar stool). But I'll relay those adventures later in a future post.

It was an easy decision for me to head out the main gates, especially after having to stay aboard the first day in port. It was a little harder decision to decide where to get a tattoo in Hong Kong, and an even harder decision to decide which tattoo I wanted, once I decided to get one.

But like most journeys in life, sometimes the path is lit for you. Not all tattoo parlors offered free beer with your tattoo. That little tidbit of information helped to narrow down the choices.

A question for today's "inkers": do tattoo parlors (are they still called that?) offer free beer with a tattoo? Asking for a friend. [<grin>]

My friend, Charlie Brown

On this "tattoo" day, I went into Hong Kong with a friend, a fellow shipmate. Charlie Brown. Yep, that was his name, "Charlie Brown." Not "Charles" and no, he didn't have a middle name; just plain ole' Charlie *NMN* Brown. He was a Yellow Shirt director in Fly 2 (amidships) on the flight deck of *Kitty Hawk*. Charlie was a great guy, kinda quiet, but one hell of an aircraft director.

You could always count on Charlie Brown to watch your back on the flight deck. More than once, I heard him over the headset yell, "Watch it, Smitty; jet blast."

That basically meant that Smitty (another great Yellow Shirt director & friend) was fixin' to get blown down by a turning jet's exhaust, something *all* flight deck personnel experience at least once in their "flight deck lifetime." I was no exception.

When you hear "jet blast," you do what all firefighters tell you to do when you're on fire, "Stop, drop, and roll." Most of the time that worked.

But in case you wanted to see what getting blown down the flight deck actually looks like, there's a short 50-second YouTube video. Don't worry. There's no gore and no blood, just a fantastic "yahoo!" finish. Note that not everyone who got blown down has an encore like this fellow.

Pinky's Tattoo Parlor

So, Charlie & I ended up at Pinky's Tattoo Parlor in Hong Kong. Charlie knew the type of tattoo he wanted. I was *really* only going with Charlie to make sure he didn't get into any trouble and do something stupid [<wink, wink>].

At the time I was there in '75, Hong Kong was still under British government rule. There were literally dozens of tattoo parlors, all within a short walk of the port. I should also note that it was against Navy regulations to get tattoos in Hong Kong because you were never sure how clean the needles were. I guess being so young and so naive, we simply didn't care.

After all, we were Navy and *we were invincible.*

Once inside Pinky's, Charlie sat down in the chair to begin his tattoo "session." I looked around at all the available designs and one of the other artists came up and asked if I wanted a tattoo.

I don't know why, but I keyed in on a particular design that stood out among the hundreds of others. He asked "Which arm?" in broken English and I was hooked.

Free Beer

I sat down in the chair and watched this guy first trace the tattoo design on a piece of tracing paper. He then cleaned my upper right arm with soap and water and slapped the tracing paper on my arm, allowing the design to transfer to my skin.

Meanwhile a sweet young thing brought me a beer. It was free. It tasted good.

"Bring me another, please."

The artist then started inking the tattoo using a needle gun with a needle that had a quick in and out motion, much like my mom's old Singer sewing machine. He would dip the needle in a color ink and then started tracing the outline with the needle gun, periodically wiping the blood off my arm. It didn't hurt too much.

"More beer, please."

I was really fascinated by this artist's ability to first tattoo in the outline, then start coloring in the other features. He told me in broken English that the first colors to fade out would be the whites, then the reds. The blues would be there until a "couple of weeks after I died."

He was not lying. As I write this today, 45 years later, the whites have all disappeared and the reds are almost totally faded. The greens and blues are still bright and crisp.

When he finished, he washed everything down with soap and water, then applied some Vaseline and put a gauze bandage on, telling me not to wash it for two days, then pull off the bandage and wash well with warm soapy water.

"Would you like another beer?" "Why, yes, I would."

A "Barber Pole"—are you f***ing kidding me?

When Charlie and I were almost done getting our tattoos, a sailor staggered into the tattoo parlor. I didn't recognize him, but then, there were more than 5,500 sailors aboard *Kitty Hawk* during the cruise.

He was obviously drunk, but staggered around and said proudly at the top of his voice, "You guys ain't got no balls." At which point, he dropped his pants and skivvies to show off his "barber pole."

Apparently, a few days earlier he had someone tattoo a barber pole onto his tallywacker. Everyone in the parlor roared. That was funny enough, but for years we've often wondered, "How did they do that?"

(Think about it… or *don't* think about it). It still boggles my mind.

To this day…

I guess I'd had about three or four beers, all free during the session. Getting the tattoo didn't hurt much and it is the only one I have.

At the time, my tattoo cost me forty Hong Kong dollars, or a little more than eight U.S. dollars.

I don't know how they do it now, but I'm sure the technology has changed a little. At least, that's what my daughter told me. I see tattoo parlors all over the place now and I see more people with more tattoos, not just veterans.

Some are, IMO, *easier* to look at than others.

I never got an infection and never wanted to have it removed. It is part of my soul.

Besides, my grandfather on my mom's side had been in the Navy and he also had a tattoo. That was how I would break the news to my mom… in a letter… I'd start off by asking if Granddaddy had a tattoo.

"BTW, Mom. It's not a barber pole… it's an American Eagle… And I'm damn proud of it."

(16) WWII—75 Years Ago... Crossing into the Siegfried Line

V-E Day (Victory in Europe) was May 8, 1945. You are probably aware that this year marks the 75th anniversary of the end of World War II. There will be many (MANY) celebrations held in Europe over the next few months. It is still a big deal to those who were liberated.

It's also a big deal to me, because my father fought during this war and was part of the liberating forces.

FYI: I'm planning to write a few blog posts and sprinkle them in over the next few months with excerpts from my WWII book: ***You Can't Get Much Closer Than This: Combat with the 80th 'Blue Ridge' Division in World War II Europe.*** This was my dad's WWII diary that I edited and published in 2005.

I think some of the passages will be appropriate to share during this 75th anniversary of the end of WWII.

While Pop is no longer with us (my dad died from cancer in 1989; he was only 68), I know he appreciates that we all continue to keep these memories alive so that future generations will *Never Forget.*

Early February 1945

The Battle of the Bulge had "officially" ended January 25, 1945. It had been the coldest winter Europe had seen in more than fifty years. Lots of snow, much of which was beginning to melt, made the rivers rise to unusually high levels.

The 80th Infantry Division, like most Army divisions who fought in Europe, crossed many rivers—it was *almost* routine. But the early spring '45 thaw made the crossings much more difficult and dangerous. And the Germans knew that.

My father's own words:

We knew that our next mission would be to cross the Sauer and Our rivers and hit the Siegfried Line (known as the West Wall), and that wasn't going to be any picnic. Although we never talked about it, we had a pretty good idea what had to be done when the time came.

We found some liquor in the house we had temporarily moved into, awaiting the crossing. My good buddy, Kad [Lt Saul Kadison], Bill [Lt Bill Mounts], and I enjoyed ourselves as much as we could because we were only a few miles from the German border.

The Siegfried Line

German engineers had constructed concrete "dragon's teeth" in parallel rows across hills and valleys. Mostly, they were five rows of pyramid-shaped, reinforced concrete projections resting on a concrete mat, ten to thirty meters wide and sunk a meter or two into the ground.

The dragon's teeth were staggered and spaced in such a manner that a tank could not drive through. Interspersed among the teeth were minefields, barbed wire, and pillboxes that were virtually impregnable by artillery and set in such a way as to give the Germans interlocking fire across the entire front.

February 11, 1945

It was cold and rainy, and there were only walls of houses remaining. That afternoon we went up to take a look at the place where we would have to cross the Sauer River into the Siegfried Line. The crossing site was in the vicinity of Dillingen, Luxembourg.

From our vantage point on the mountain, we looked down to our right front and could see the battered little town of Lellingen, where the engineers were going to try and build a Bailey Bridge. The surrounding ground was nothing but a pocket of shell craters, and the Krauts were still firing artillery.

The Sauer River was swollen many feet above its normal level and the current was very swift. It didn't look good at all. The river wasn't normally that wide, but an unusually early warm spell had thawed the snow, causing the river to swell to about twice its normal size and flow at about ten miles per hour.

We were going across in assault boats. One platoon had already made it over and had established a perimeter defense. The only things that should bother us were artillery and snipers. The Krauts were sending down combat patrols every now and then, but they didn't amount to much.

My mortar platoon was to be the last across. While we were in the CP [Command Post], one of our men came in soaking wet. He had gone over on an engineer rubber raft. The current was so strong that it tore the raft to pieces and several of his men had drowned.

The engineer captain who was in charge of the boats had a cable stretched across the river and had a system rigged up where the current would pull across the assault boats and rafts. It was not the first time we had crossed rivers this way. But it was the first time we tried to cross a river flowing as fast as the Sauer River was that day.

Time to Cross—Now!

About midnight, on my way back to rejoin the platoon, I met bringing the platoon down the road. He had received orders to get the men across right away. The current was very swift, he advised, and we could only put three men with their equipment, plus the engineer in charge, in each boat.

It was pitch black and we couldn't see a thing. Something didn't feel right, and I was getting nervous. We had crossed rivers before at night in assault boats and rafts, but this was different. I couldn't explain it.

Mac (my runner), Sergeant Roach, and I went across first. Kad was going to feed the men to us. We were going to assemble on the opposite bank and then move west about half a mile to where Bill was positioned and waiting for us.

Two boatloads of my men made it across before the boats stopped running. There was no way we could get in touch with the other side to find out what was holding everything up.

I went back to the wire cable, where I had left Sergeant Roach, but no one had come across. I sat down on the bank of the river with my hand on the cable.

That way I could tell when a boat was coming across.

The cable grew taut. A boat was coming over. The cable grew tighter, then it snapped.

It was an awful thing to witness. The night was so black that we couldn't see our hands in front of our faces, and yet we could hear people yelling for help only a little way out in the river. We were helpless. If we tried to help them, we would surely have drowned ourselves. We didn't have any lights with us and even if we had, we probably still couldn't see anything.

Where's my buddy, Kad?

When dawn came, we got the rest of the men across the river. I knew that Kad would see that all the men got across before he started, so I didn't ask anyone where he was until Sergeant Peck came across and asked me if Lieutenant Kadison was okay. I told him I hadn't seen him.

Then Peck told me that after the boats broke down, Kad crossed on a raft to tell me what the hold-up was. He was probably on the raft that had been torn apart in midstream. My heart sank.

It wasn't a pretty picture. Kad had on his overcoat and all his equipment. I could picture a flimsy raft tearing apart in midstream of a swollen river and the hundred boxes of heavy rations tumbling every place. It was so dark that visibility was impossible. Even an excellent swimmer could not have gotten out of that death trap.

I tried to believe that he had somehow escaped and was somewhere down the river. Nothing could happen to my buddy, Kad. He had escaped death so many times that no little river could hurt him. We had shared so much together. The snows were gone now and we would soon be through the Siegfried Line. The war would soon be over and we could relax and enjoy life together. He was a brilliant man and had so much to live for. I loved him as my own brother, and he meant so much to his men.

I did everything possible to find him. We asked everyone we saw. We checked every hospital and every aid station, asked every medic. The 5th Infantry Division was crossing down the river a piece. We asked everyone there. Sometimes we got rumors. We traced each of them down, but they only ended in blind alleys. One man told us that a dark-haired man who had been wearing a red scarf had been picked up out of the river. Kad wore a red scarf, so I got my hopes up. But that led to nothing.

The Krauts had been sending combat patrols down to the river to try and get some of the prisoners back. Perhaps they had gotten Kad. If that was the case, then he was sure to be all right.

But that was not the case. His body was found two months later, three miles below where we had crossed. Lieutenant Saul Kadison from Brooklyn, my friend, my buddy, my brother was dead.

Emotional Scars

My father *never* talked to me about his friend, Kad. I first read about him in my dad's diary. Even then, he wouldn't talk about it—he would always change the subject.

Losing a fellow soldier, especially one that became one of your best friends under the most extreme stressful times, is one of the most difficult things a veteran carries with him throughout the rest of his/her life.

Books are written about the "band of brothers," but I don't think any of them can convey the sense of closeness veterans have with other veterans, especially those they served with.

My dad made it through the war with a few *physical* scars. He was awarded two Purple Hearts and a Bronze Star Medal for Valor. I know he carried a lot more emotional scars, but his generation rarely talked about the war.

More than once I told him that he was my hero. I know, deep down inside, he appreciated that, but he always told me, "Son, I'm not a hero. The heroes are the ones who didn't make it back." I'm sure many of you have heard that exact same statement, too. And not just in the movies.

I think all of us appreciate what our fathers and mothers (*and* grandfathers and grandmothers) sacrificed during World War II. I'm of the "Vietnam Era" generation and I, too, appreciate what ALL veterans have sacrificed. I thank the good Lord that I never saw combat, though I did see my share of death during my time in the Navy.

This is one of the most somber and powerful statements about veterans I've ever read. I'm sure you've seen it before and appreciate it like I do. If not, please read carefully.

Whether active duty, discharged, retired, or reserve, a Veteran is someone who at one point, wrote a blank check made payable to "The United States of America" for an amount of "up to, and including, their life."

THAT IS HONOR
We Will ***Never Forget***

(17) From Sailor to Civilian

Let me take you back in time a few years. We all remember… it was just a *few* short years ago when you left the Navy and became a civilian again.

Remember when you joined the Navy (Army, USMC, USAF, or USCG)? Boot camp, "A" School, your first duty station? Your other duty stations, cruises, shipyards?

Yeah, you remember *most* all of those milestone events, don't you? I wrote an earlier post about *my* journey from "Civilian to the Navy."

How many of you can recall your first day (or first week) *after* you were discharged? Where were you (your last duty station) and how long did it take you to get home?

Many of my fellow shipmates were married and I know their wives & families were anxiously waiting for them as they walked down the gangplank and into the arms of their loved ones. Like you,

I witnessed many those celebrations almost every time we returned back to our San Diego home port.

It was especially emotional when we returned from our six-month Westpac '75 **cruise.** I know many of you remember those special reunions. It's something one never forgets.

The DD-214—Freedom!

Me? I was on *USS Kitty Hawk* and was honorably discharged from the Navy on July 8, 1977 after four years of military service. Actually, it was ***Three Years, Eleven Months, & 29 Days—But Who's Counting.***

It took a while to process out. Personnel used typewriters for all the forms with carbon paper. They didn't use computers back then. But I had all my paperwork, including my most important **DD-214** discharge papers ("press hard, the third copy is yours!").

I had emptied my coffin locker and my coat locker—and my "secret" stash locker in Crash—and packed up my sea bag for the last time. I wore civilian clothes—blue jeans and a pull over shirt. I made that trip one more time from the V-1 Division berthing compartment on the 03 level, down the stairs, across the hangar deck, and the lowered #3 elevator to the after brow.

As usual, I saluted the petty officer of the watch and asked one more time, "Permission to go ashore, sir." He bid me farewell and good luck. I saluted the flag one more time, even in civilian clothes—an old, *continuous* habit.

Leaving one home for another

I had a plethora of mixed emotions. Unlike when I left NAS Agana, Guam, I wasn't leaving a paradise—I was *leaving a home.* I wasn't transferring to another duty station—I was leaving *for* home.

Kitty Hawk had been a great ship to me (and my "home" for 2½ years) and most of the crew that I worked with I considered family.

There was no fanfare, no band, and no one on the pier waiting for me. There were no tears or sorrowful looks. At the time, I was ready to leave *Kitty Hawk* and the Navy and put both far behind me. It was time to move forward with my life, though I had no friggin' clue what that would be.

With that simple gesture, I was no longer a member of Uncle Sam's Sailing Club. I was once again, a civilian. Of course, at that time it didn't really feel any different and it would take some time before I realized I was truly out of the Navy.

My sister, Anne, had flown out to San Diego from Florida. We were going to drive back to Florida together and see the country. We took our time and it was a great trip. While I had seen her from time to time when I was home on leave, we had both changed—I like to think *matured.* We enjoyed getting to know each other again. Nothing like having an older sister to visit and catch up with during a long drive home.

Coming Home

When I got back to my mom & dad's in Florida, it was another great homecoming. I knew I didn't have to leave anytime again, so I was home—really *home.*

Or… so I thought.

My mom was a great cook. My dad, *well…* he had his "specialties." I'll spare you the details, but it was great to have home cooked meals again. *Some* things just didn't change.

My plan was to go back and finish college. Before joining the Navy, I'd spent a year at Florida State University. They'd asked me to leave after my first year—something about my 0.7 GPA. So, I decided I wasn't going back to FSU. Instead, I was going to attend the local community college and first, get an Associate's Degree.

When *Kitty Hawk* was in the Bremerton Shipyards in 1976, I had taken an electronics course at Olympic Community College. I really enjoyed the course and did well (nothing like those C's & D's @ FSU). From that simple one-semester experience, I decided I was going to be an electronics engineer—that was *my* goal.

So I knew I'd need a lot of college. But that wasn't a problem. I had plenty of time now.

City Firefighter? Why not.

I also needed a job, but I wasn't sure what to do. However, another opportunity ("journey") presented itself. I joined the Gainesville Fire Department and worked full-time—1 day on, 2 days off. Since I'd been an ABH (Aviation Boatswain's Mate-Handler) and spent the majority of my four years in Crash, that seemed like a great transition. I figured I could work full-time and attend college full-time. *Easy, peasy.* Yeah, right.

That worked… but only for a little while. After 2 years, I graduated from Santa Fe Community College with an AS degree and started full-time at the University of Florida. My GPA took a nose dive from 3.6 to a 2.9—ouch!

I figured I needed to concentrate on college, so I resigned from the fire department and long, story short—I graduated with a BSEE in 1982. And most important, I met the love of life (in a Karate class of all places) and married in 1981. We're coming up on 39 years together!

Being an electronics engineer was my *third* career ("journey"). But I was on my way to bigger & better things. More about that later.

I wanted to mention a couple of other things that happened along the way. Perhaps you may have also experienced similar "events" when you arrived home.

Not the Same Home I Left

I left home (the first time) when I was 17 years old. I spent a year at FSU, then I joined the Navy. After four years, I returned home.

The last time I "lived" at home was five years earlier when I was 17, fresh out of high school. My mom & dad were somewhat strict parents (I loved them dearly, but you know what I mean). So when I returned home *after* the Navy, I think they thought the "rules" were still the same. You know… like when I was still in high school and 17.

Well, *my* rules weren't the same as *their* rules…

Stay with me here, you know where I'm going with this.

In the Navy… when in port… I often stayed out until the early morning hours—no surprise. I'm sure many of you did the same, especially when out on a cruise.

Now, back at home, sometimes, I came home in the middle of the night. Sometimes, I stayed over at a friend's house. *I* didn't have a problem with that.

But my mother...

Something's Gotta Give

Well, let's put it this way. One morning, I *casually* strolled in while my mom & my dad were having coffee at the kitchen table. I said, "Good morning," with a bright smile. But instead of responding, my mom got up from the table, got dressed, and headed out the door to work, without even a "Hello."

I asked my dad what was going on—he had fought in WWII and "knew" I was enjoying life after the military. I think *he* was okay with me staying out partying.

He wasn't taking sides, but *kindly* suggested that if I was going to stay out all night, to "at least have the courtesy to call and let them know that I wouldn't be coming home that night."

Well, I think it was either the second or third phone call at one o'clock in the morning that he put an end to that crap.

We... came to an agreement *of sorts.*

Since I had just started working at the fire department, it was time to move into my own place. That wasn't a problem—I was making $9,600 a year as a city firefighter (big money back in 1977) and I rented a brand-new studio apartment on the other side of town for $350 a month.

After that, we all got along great. My folks didn't have to worry about me anymore.

Strange how that works, isn't it? Of course, when *our* own children were growing up, they probably thought *our* "rules" were strict. But that's another story for another day.

One more...

Did you see the movie, *The Hurt Locker*, starring Jeremy Renner as a member of an Army bomb squad unit? Yeah, *that* one. It won six Oscars.

There's a scene in that movie when SSGT William James (Renner) is first discharged and happy to be home with his wife. They go grocery shopping. She's got a full grocery cart and asks him to pick up some cereal—"meet you at the cashier."

The camera shot is perfect… a long distance away, we see him standing in the aisle, all alone, with shelves and shelves of cereal boxes—too many decisions. This is a man who'd been making instantaneous life & death decisions every day for the past year. Too many friggin' choices in his new civilian life.

I wasn't EOD (though I did test for it in San Diego), and I'm *certainly* not Jeremy Renner. But for a sailor who'd been in a certain routine for years on end, I kind of felt that way when I first got home—I felt out of place.

Different Routines

How am I supposed to act & react to situations that I personally had no control over? I didn't have that big a problem adjusting back to civilian life, but you know what I mean.

I'd been getting up at the crack of dawn to the lovely tune of *Reveille* for several years. I'd stood watch at two in the morning watching several of my drunk sailor friends stagger up the gangplank, swaying side to side, with a big ol' grin on their face from ear to ear.

My showers were three minutes long, *including* shampooing my gorgeous crew cut. My head was always on a swivel on the flight deck, watching for any danger and all movement from every angle you could imagine.

I simply didn't have to do that crap anymore. Some of those routines were hard to drop; others, not so bad.

But to this day, I still keep my head on a swivel, especially when riding my bike. My wife & I taught our kids to always (ALWAYS) be aware of their surroundings and look confident wherever they are. When there's a car wreck or a fire, I still run *towards* it to see if I can help.

Three minute showers? Not so much these days. Unless of course, we have a hurricane that knocks the power out for a week. Then, that old Navy routine is handy.

But I digress…

The Navy *Never* Left Me

While I've been out of the Navy for almost 43 years, there are still some routines and traditions I continue to follow.

I still fold my skivvies the same way. I still tuck my bed sheets in with that 45 degree fold, though I haven't tried to bounce a quarter off of it to test how tight they are. I still check my "gig line" when I wear a button-down shirt.

And yes, I still use a lot of the same Navy language I picked up in my "career." Of course, I try [<wink, wink>] to be cognizant when I use it and in most cases, it's appropriate. At least, I think so.

I get teary-eyed at some TV commercials, especially those involving our veterans returning home. I sometimes find myself having to take deep breaths when I see or hear or smell something that takes me back to the flight deck. Most of the time, they are awesome memories and I am grateful for that. I rarely have negative thoughts about the Navy. After all, isn't that how we continue to move forward?

I'm grateful to be a veteran and I respect *all* veterans of all military branches: those who served before us, those who served with us, and those who serve now (our future veterans). I hope you do, too.

Some of us had an easier time "coming home." I am one of those. I hope and pray that you feel welcomed, wherever you are in this world and wherever you are in your lives.

And you know what else? I wouldn't trade my time in the Navy for anything. I am glad to be where I am now—a *civilian*. But I can never forget that I'm a veteran—a proud Navy veteran.

(18) Things to Say During My Exit Interview

I've had several great careers and in some sense (even though I'm "retired"), I feel I'm still on that journey for the next one.

Each one seemed to build upon the previous and prepare me for the next (*Life is a journey, one step followed by another…*). I don't know if that was skill, luck, or fate. Whatever, I am one very blessed man.

But my first "career"… Navy! I truly believe it set my life on the right track. You know: teamwork, leadership, ability to BS with the best… Yeah. *That* Navy.

However, after I left a particular job, I sometimes wondered if I was making the right decision, doing the right thing… you know what I mean. Leaving a sure thing (good *or* bad) for an unknown.

It's not like transitioning from a Sailor to Civilian, though. It's something more, something deeper, something bigger. Yet at times, I have looked back and asked myself, "What could/should I have said during that exit interview"? I'm sure some of you may have also experienced similar thoughts.

The Decision to Move…

Several years ago, I was hired as the Chief Information Officer for a large law firm in West Virginia. I'd been a legal technology consultant for 25+ years and traveled around the country working with more than 300 law firms, law departments, law schools and government agencies of all sizes and types.

Every now and then, out of the blue, one of the senior partners or law firm administrator would ask if I would be interested in coming to work as their IT Director or CIO. Chicago, New York, Dallas, Charlotte?

Nope, I liked being a consultant. I'd share this with my wife, Becky, and we'd think for a moment… a quick moment… and agreed, "Thanks, but no thanks. We like it here in Florida."

Imagine the dinner conversation when I came home from a consulting gig and asked if she wanted to move to West Virginia.

Well… I'll spare you the details, but our kids had moved out, graduated from college, and were off the payroll. My parents had passed away and Becky's were happily living in a retirement community.

Yet… I'd always wanted to live in a place with seasons.

Let me pause here for a quick moment and share with you that Florida has two seasons—green and brown.

So to make a long story short, I accepted the position and we moved to West Virginia for four years. It was a great law firm, a great job and we both loved living in West Virginia. It is one of the most beautiful states I've ever been in.

And before you ask, I have visited all 50 states, plus Puerto Rico and Guam (I was stationed at NAS Agana, Guam for 15 months during my time in the Navy).

How "The List" Started

When you're in a new leadership role (in my case, a large law firm), you have to work to establish yourself as both a leader and a visionary. It doesn't just happen—at least, not for me.

It's not like in the Navy when one day you're an E-3 Airman and the next you're a Third Class Petty Officer… now in charge of a group of men who a few minutes before were your peers and your friends.

You have to develop relationships at the executive level, the law firm level, the department level, each office (the firm had 10 offices when I started), and the individual level. In other words, the people you work *for* and the people you work *with*.

Even when you arrive with a solid set of credentials, you are initially under a microscope at so many levels. There are a lot of expectations—especially within your own department—and you need to build trust and confidence and look for various ways to do that.

Keep in mind that most of the folks in the IT Department had been there for years, were used to the "firm's culture," and their own "IT culture."

I'm okay with that; it's simply part of the process.

This "exercise" initially started as a weak attempt at humor with some of the folks I worked with in the IT department. In my mind, it ended up as a continuous team-building exercise and better yet, some of these suggestions *came* from my staff.

#1 on the List of Things to Say During My Exit Interview

After my honeymoon period was over at 3 months, I made a comment to one of my managers, "It sounded like a good idea at the time." To be honest (see "C.R.S." below), I can't remember the context. I wasn't (I don't think) in danger of being fired, and we all laughed about it—more than once.

But someone suggested that should be on "the list of things to say during my exit interview."

For those of you who run an IT Department, you know what I'm talking about. For those of you who *don't* run an IT Department, But actually have a *real life* elsewhere… you probably also know what I'm talking about: It sounded like a good idea at the time.

Note: this is not the same as, "Here, hold my beer." I'll share those stories in a later post.

#2 on The List of Things to Say…

Some CIOs use the "iron fist" method—"my way or the highway"—and others take a more subtle approach. A lot of this type attitude has to deal with the culture of the particular law firm and even within the IT department. Most important, your own personality. People are smart and can see right through the superfluous BS.

When I first made the comment above, one of the IT managers quickly retorted with, "You should have asked for more money," which became item number two on the list of things to say during my exit interview.

Growing The List…

And so, The List began to grow, mostly by me, but a few by my staff. I looked at this as a way of building relationships within the IT department. For the most part, it was all good humor.

I actually had this list on the wall in my office, for anyone to see, even the Managing Partner. More than once I'd walk into my office to see that someone "penciled in" another item to the list. For a while, I wondered if they were suggesting I should leave or if they were simply having fun.

Since I was there for four years, I'd like to think the latter.

The Rest of The List

It depends on the day and the situation, but I'm willing to bet that most CIOs and IT Directors and others in leadership roles have uttered some of these to themselves more than once. And now that I think about it, these could relate to anyone… anytime… anywhere.

They are not in any particular order.

"I should have taken the blue pill"—this almost became the mantra by some IT members. If you remember the movie, "*The Matrix*."

"*Release the Kracken*"—we used to say this right before a major technology rollout, or after I had a confrontation or difficult meeting, especially with a higher up.

"*The FM Box broke*"—for those of you who don't know about the *FM Box,* to the normal computer user, there are inputs and there are outputs. In between—*in the Box*—the magic happens. Hence, the *FM Box*. What does FM stand for? Google it, "FM Box." Hint: something close to, "the *freaking* magic box."

"*IETSU*"—I eat this *stuff* up. In the Navy, we used to call this "Living the Dream" in response to when someone asked you, "How's your day?"

Speaking of the Navy, another quote on the way out the door, "It's not just a job, it's an adventure."

"*BTSOOM*"—beats the *crap* out of me. "I'm not the Help Desk,

I'm the CIO. I don't know how the thing-a-ma-jig works—I have people that deal with that."

"*I brought you into this technology world—I can take you out.*" I usually uttered this phrase under my breath after one of the attorneys showed their "unprofessional" side. Yeah… that crap happens outside the Navy, too.

One of my favorites, especially since I recently celebrated the 28th anniversary of my 37th birthday: "I've been diagnosed with the CRS Syndrome"—Can't Remember *Stuff*.

"*Was Shakespeare in IT*?"—OK, I know this will stump some of you non-theatrical people—Google "Shakespeare Henry VI quotes" and keep in mind, I worked at a law firm.

The Last on the list of things to say during my exit interview

"Thanks for the ride. I've enjoyed my time here and I hope that I've been able to contribute moving the firm a little farther down the technology track. My best to you for the future! Thank you and may God bless!"

(19) The 3-Minute Navy Shower: A Luxury, Compared to…

"Survival is one of the most basic instincts of mankind."

Somebody smart wrote that once. I know it's true and it was from a long, long time ago because I found it on the Internet. It's by someone named "Anonymous."

Well… "Anonymous" was probably a Third Class Petty Officer in the United States Navy. Or, perhaps a Corporal in the US Army or USMC. Or an airman in the USAF. I can't remember exactly, but I know s/he had to have served in the military at one point. I mean, what a profound statement, right?

Learning to live on an aircraft carrier with 5,500 of your *closest* friends… out at sea… on a 6-8 month cruise is… well, a survival technique.

Imagine, if you would…

There were about 80 of us in the V-1 Division aboard *Kitty Hawk* (CV-63). Back in my day (mid-70s), there were no female sailors on board.

We were a mixed bag, too. I mean that in the kindest way. We came from all corners of the country, we were all different sizes and shapes, and we all had different cultural backgrounds. Each of us grew up ("survived") in different environments and learned to work together as a close-knit team on the flight deck of an aircraft carrier.

We also had to live together.

When *Reveille* sounded at 0600, we popped out of our racks (beds) and migrated toward the head (bathroom)… all 80 of us; some moved quicker than others. On *Kitty Hawk*, there were four shower stalls in our berthing compartment.

The 3-minute Navy shower

Let me introduce you to the *3-minute Navy shower*—it may help you survive the next time power goes out and there are several of you living under the same roof.

Stay with me here, this is about to get interesting.

Step one: step into the shower stall and pull the plastic curtain for privacy. If you've ever been to Italy or Greece and stayed in a "non-luxurious" hotel, well… you know the size of the shower stall I'm talking about.

Step two: turn on the water. *Kitty Hawk* showers had spring-loaded, automatic off control knobs. They'd automatically shut the off water as soon as you let go of the handle.

Step three (assuming you figured out Step two): the first time you take a 3-minute Navy shower on a ship with these auto-off knobs, you quickly learn that the hot water is REALLY hot and the cold water is REALLY cold.

Step four: you cuss under your breath and suddenly realize that the Navy figured out a nifty way to save water. And that's not a bad thing when you're at sea for weeks at a time and the "fresh" water is actually made from salt water and… it takes time.

But I digress…

Water Hours

There were times during the Westpac '75 cruise, especially when we were out more than a couple of weeks, when there would be NO HOT WATER available and we had to take COLD showers. That is, if the water was still turned on. In all it's wisdom, the Navy also figured out if the water was really COLD, you probably wouldn't stay in the shower for long. And… they were right!

Step five: After you get under the shower and get that first drench, you can let go of the control to turn off the water. The next thing you do is to quickly soap up, then quickly rinse off. All the while, trying to keep your hand on the shower control while you rinse those hard to get places.

So, that's the gist of the 3-minute Navy shower. Quick and simple. And when you're done, you step out to dry off while the next guy jumps in.

What about… rain?

Picture this…

You're in the South China Sea… it's July… you're only a few miles from the equator. It's hot and it's humid, but you're working on the flight deck of an aircraft carrier. You're wearing long pants, long sleeves, steel-toed boots, a float coat (inflatable life preserver), and a cranial helmet. And… you're downwind of a dozen jet engine exhausts.

Sure, there's a "gentle ocean breeze" flowing over the bow… blowing that hot jet exhaust your way.

You've been on the flight deck for hours. You're dirty, you're sweaty, you're greasy, and you have spatters of JP-5 jet fuel all over you. You didn't get a shower the night before because you got off the flight deck long after water hours closed the showers.

It starts to rain. What do you do?

For those of us on the flight deck, if we were in between flight ops, we'd simply strip off our shirts and let the rain flow, baby! A gift from God for the hot & sweaty.

I still have those fleeting thoughts here in Florida during the hot & humid summer days when it starts raining. My wife & kids have more than once caught me outside, head tilted toward the sky, shirt off, taking in the rain. It's one of those memories I'll NEVER FORGET. "Livin' the dream, baby!"

Lessons Learned

The lessons to learn here are two-fold: #1, be quick and #2, be courteous and respectful.

While #1 no longer applies anymore (at least to showers), I do still try to follow #2—*be courteous.* It's one of those things I learned early in life and was affirmed time and time again while in the Navy—some things just stay with you.

I'm willing to bet that most Navy veterans rarely take a 3-minute Navy shower now. Face it, we're at home and we have luxury of taking long (LONG) showers if we want to. And why not? We endured years of those short showers, so we deserve *our* time now.

I say that jokingly, but there are others in the military who *were* not as fortunate as us sailors. I have a lot of friends who served in the Marines and the Army. While us Navy vets may have "complained" about the 3-minute Navy shower, let me share with you another story.

For my Army & Marine friends

My dad fought in WWII with the 80th Division in Europe. He told me one time that he went 39 days without a bath. Wow! That was on the extreme, I know, but I'd also heard similar stories from other WWII AND Vietnam veterans.

My dad was wounded in October 1944 during an attack at Sivry, France. He was on the second floor of an abandoned house, acting as a forward observer for his mortar platoon when it was hit by German artillery. The roof caved in on him, knocked him out, and pinned his knee, tearing the cartilage.

He ended up going from the battlefield, to battalion aid (about a mile from the front lines), to the collecting company, to the Evacuation Hospital in Nancy, France. He eventually ended up in England for surgery and to recover from his injury, but *this* story is when he first reached the Evac Hospital in Nancy.

This is an excerpt from my WWII book, ***You Can't Get Much Closer Than This: Combat with the 80th 'Blue Ridge' Division in World War II Europe***. These are my dad's words…

I must have been quite a sight to behold. I hadn't shaved in several weeks and hadn't brushed my teeth in quite a while. When I looked at my teeth in a mirror, they were black!

The nurse said I needed to check my valuables, so I emptied all my pockets. I put everything in a pile beside me: my wallet, K-rations, cigarettes, toilet paper, gum, can opener, a leather thong I used to tie up my raincoat, a can of powdered coffee, some dirty oily rags I had been using as handkerchiefs, and a pair of binoculars I still had around my neck.

In my left-hand fatigue jacket pocket I had a .45 Colt pistol that I always carried in case my carbine failed. I had even forgotten about that. In addition, I had a few little souvenirs I had taken from some prisoners we had captured: a piece of silk from a parachute and some emblems and patches ripped off Kraut uniforms.

After the pocket inventory came the process of undressing me. That was rather embarrassing because the last time I had a bath of any nature was out of a helmet thirty-nine days earlier, and except for my socks, had not changed clothes. I hadn't had a hot bath since I left the States [July '44]. Naturally, I couldn't smell myself, but I know the nurse could detect a distinctive barnyard aroma about me.

Nevertheless, she was an understanding American gal and didn't make any comments. I was wearing a field jacket, fatigue shirt, fatigue pants, OD [Olive Drab] pants, two OD shirts, two wool undershirts, wool drawers, cotton shorts, and a cotton undershirt. All of them, particularly my underclothes, were black as the night and almost stiff enough to stand up by themselves.

My cotton shorts still had a hole in them where a piece of shrapnel had come through them several weeks earlier. There was still dried blood on them. The nurse helped me down to the last mentioned item and then the ward boy took over.

My body was black with dirt. I was on the way to the most wonderful experience I'd had in quite some time—a hot bath! The ward boy wanted to bathe me in bed, but I said, "Hell, no. You just help me to the shower room and

I'll take over from there." I bribed him into carrying me to the showers. My leg hurt like hell, but that shower felt mighty good.

I have NOTHING to complain about

After two months, my dad made it back to his unit (317th Infantry) on December 14, 1944—two days *before* the Battle of the Bulge began. He survived the war (obviously, or you wouldn't be reading this blog) and made a great life for himself as a husband, father, and lawyer. Unfortunately, we lost my dad to cancer in 1989—three days shy of his 69th birthday.

My mom once told me that when they first met and eventually married, my dad was taking two to three showers every day. I wonder why?

Never again will I complain about the 3-minute Navy shower.

(20) Port Call: Philippines

"It's not just a job. It's an adventure!" Who remembers the first time stepping into a *new* country as a member of the United States military?

No, I'm not talking about *boot camp* or *Honolulu*… I'm talking about a country *outside* the United States. And it doesn't matter if you were Navy, Army, Marine, Air Force, or Coast Guard.

It's one of those early military memories I think we all remember. For some of us, it was a pleasant memory.

But I know there a lot of veterans would rather forget their first step in-country, especially Vietnam. I was fortunate enough to miss Vietnam—I joined the Navy in July 1973. I also realize that not everyone was stationed *outside* the U.S. But stay with me here…

For me? It was the Philippines. We always called it "PI," short for the Philippine Islands. It was during the Westpac '75 cruise—my first and my only Navy cruise. I have a lot of great memories of my time in PI.

Subic Bay Naval Station was our overseas home port. It became the U.S. Seventh Fleet forward base for "repair & replenishment" after the Gulf of Tonkin incident in 1964 (i.e., *The Vietnam War*). It was closed in 1992.

When I was there in '75, we were in and out of the Philippines a total of seven times.

So How Would I Describe PI?

San Miguel beer, baluts, Jeepneys, grilled monkey meat, *The River*, and martial law. Did I mention *cheap beer?* Yeah… you remember *that*, don't you?

Think back on *your* first duty station outside the U.S. How would *you* describe it?

Go ahead… I'll wait.

Baluts—how do you describe a balut? I may insult a few folks here—it is *not* my intent. But think of the most disgusting thing you've ever encountered and double down. Take a fertilized chicken egg, bury it in the hot sand on the beach and leave it for about three weeks. Dig it up, cook it, remove the top of the egg, and eat it… Yeah, you heard that right.

It was considered a delicacy by Filipinos. But I can honestly tell you after seeing of one those things up close (and *almost* personal), I didn't need to see any more. I think I'll just leave it at that.

Question for my Navy friends. How many of you ate one of these when you were *completely sober*? Honest answers only.

Jeepneys were old WWII jeeps that had been refurbished by Filipinos. They still had the original engines, but had extended open rear end cabs—enough to carry six people comfortably or eight people cramped. Most were very flamboyantly decorated with different colors, flowers, and whatever else their owners could put on them.

Jeepneys were everywhere and after a while, you knew who drove which one by their decorations and music. They were *the* most common method for public transportation in and around Olongapo City.

Grilled monkey meat or *dog* meat? I kid you not. There were usually a dozen or so sidewalk portable BBQs of grilled meat between the Naval Station main gate and *Olongapo City* and there was no telling what they were cooking.

At first, I thought it was chicken; doesn't everything taste like chicken? But several of the older salts told me that it could just as well be monkey meat or dog meat. Who knew? Most of the time it smelled great and tasted even better. Of course, when you're half-smashed, *everything* smells good and tastes great.

Everything was Cheap

The currency in PI was a piso (pronounced "pay-so"). The exchange rate back in '75 was six pisos for one U.S. dollar. But if you knew the right people, you might be able to find someone to exchange seven pisos for a dollar.

Everything was cheap in the Philippines. San Miguel beer, made right there in the Philippines, was usually three pisos and a decent meal was usually less than fifteen pisos.

My first time in the Philippines, I went with a couple of buddies into Olongapo City, which was located outside the main gates, across "*The River*." They had been there before and I figured I would hang out with them to get an idea of what to expect and hopefully, stay out of trouble.

The streets were pretty dingy and the sidewalks were lined with wall-to-wall bars and restaurants. Almost every bar had a live band (some better than others) and there were streets with rock bands, streets with soul bands, and streets with country bands.

I spent most of my time listening to country music in a bar called the "Country Caravan." Man, could they belt out my favorite Hank, Merle, Loretta, Tammy, and Willie tunes.

Years later I would tour Bourbon Street in New Orleans and I'd have flashbacks of Olongapo City.

Outside Olongapo

Several guys in the V-1 Division were Filipino. John Tubao was the division Yeoman. He had relatives who owned several bars and restaurants in the surrounding area.

One day, he made arrangements for several of us to visit one of his family's establishments. LCDR Leone (V-1 Division CO) went with us on this trip. John managed to talk one of the Jeepney drivers down to a fairly reasonable rate—John knew how to work the system.

This particular restaurant was about thirty minutes outside Olongapo City and was located on the beach. A very serene and peaceful setting. We caught a magnificent sunset, too.

We had a complete and delicious meal with different kinds of local fish, seafood, fried rice, chow mein, special salads, and a *lot* of beer. All of his family members we met were so kind and generous—great hosts. They didn't speak English and we couldn't speak Filipino, so John had to translate for us.

Everything was tasty but… at *this* point in time and *this* particular place, I learned the most important lesson of eating in a foreign country—never, *never* (let me repeat, *never!*) ask the name of what you are eating.

If it tastes good, eat it and enjoy it. Never ask what it is, because in a foreign country, what you *think* you taste and what you *actually* eat are often two different things. I still abide by that lesson, even today.

Shore Patrol

On several occasions while in port, I had the grand opportunity to serve Shore Patrol duty. Back in my day, this fun little exercise usually consisted of teams of two sailors dressed in their whites with an SP armband and a white web belt carrying a wooden baton. For most of my SP duties, my friend, Ed Boes, also V-1 Division, was my beat partner.

Despite what you hear, SP does not stand for "Stud Patrol." SP duty usually started after work hours at 1530 (3:30 p.m.). We would walk an assigned beat, usually a mile or two in length. We'd walk into every bar and restaurant on our beat, checking to see if there were any sailors "out of hand." Then we'd move on to the next one.

I can't recall ever finding any sailors or Marines "out of hand." Drunk as a skunk, maybe, but never out of hand.

I didn't mind Shore Patrol duty and I had the opportunity to visit all the bars and restaurants on the beat—completely *sober*.

One of the more positive aspects about Shore Patrol duty was that when chow time came, we would place an order at a restaurant, and then walk the beat. When we returned to that restaurant, our chow would be ready for us—hot and tasty.

My favorite meal in PI was *Shanghai Fried Rice* and *Lumpia*. I've never been able to find it like that in the US. I'd also order a side of *Pandesal* bread—THE best rolls on the planet—served straight from the oven in a paper bag.

The food costs in Olongapo City were more than reasonable. One of my other favorites was fried shrimp. In 1975, the cost of a fried shrimp dinner, complete with fried rice and a Coke, was about $1.50. In the states, it would have cost about $3.50.

PI Martial Law

At the time I was in PI, martial law was in effect and the streets had to be clear by midnight. The Philippines was under martial law from 1972 to 1981 under the Philippines president, Ferdinand Marcos.

We usually started letting everyone know about curfew around 2330 (11:30 p.m.) and most everyone either went back to the base or followed a hooker (i.e., *Navy "Relief"*) to their home or to a hotel.

I had heard though, that the local Filipino police had orders to shoot on sight if they found anyone in violation of martial law, but I never heard of anyone getting shot.

We'd heard of some locals chasing a sailor or two back after curfew and since the base gates closed at midnight, the sailor had to swim *The River* to escape. I tell you, the decision to make: *The River* or jail—I'm not sure what I would do in that case, either.

If I Had the Chance…

Someone asked me one time what I would have done different, had I stayed in the Navy and taken another Westpac cruise to PI. I had to think about that for a moment or two.

First of all, I'd be the "old salt" and mentor to the FNGs. Second, I think I'd try to spend a little less time in the bars and more time outside the city, playing tourist. I know PI has a lot more to offer than restaurants and bars, even back in the '70s.

Of course, I'd still visit some of the old haunts, but I'd also take time to learn a little more about this beautiful country, their language, and their culture.

How about you? What would you do different if you had the opportunity to go back a second time to your first port or duty station?

(21) Reveille… and Other "Alarm Clocks" Over the Years

Reveille! Reveille! All hands heave out and trice up. Give a clean sweep down fore and aft. The smoking lamp is lit in all authorized spaces.

For ***Three Years, Eleven Months, and 29 Days***, that was my wake-up call. Not a bad way to wake up, especially when you're a young whippersnapper who joined the US Navy to "see the world."

That got me to thinking about the different methods we (actually, me, since I'm the one writing this blog post) used to wake up through the years.

I'm sure you remember most of these.

The "Early School" Years *Reveille*

Back before the Navy, before my *first* attempt at college, before high school… well, way back when I was in elementary school, "Reveille" was a little different.

You remember *those* days? Yeah, I know you do.

Remember our "Reveille" was usually "mom," with that gentle, soothing, motherly voice: "Wake up, son. Rise and shine. Time to get ready for school. Breakfast is waiting…"

Then, when we graduated to *junior* high school and *high* school, Reveille wasn't as "gentle."

My wife told me her father used to say in his *commanding* fatherly tone, "I want to see two feet on the floor in two seconds." So, Becky would slide her legs over the side of the bed and put both feet on the floor, but remained in bed… asleep.

Even though her dad never admitted it to me, I *know* he thought that was funny.

The "Working Years" (*before* the Navy) *Reveille*

Once we were out in the working world, *before the Navy*, we had this magical mechanical device called an alarm clock.

Back in my day, these were wind-up clocks. You'd manually turn the alarm to set the time you wanted to wake up. None of this AM/PM crap to figure out—*these* clocks were only 12 hours.

Just when you were in the middle of a great dream, RINGGGGGGG—that clapper bell would slap back and forth faster than a woodpecker's head on a dead pine tree. You'd reach over and either slap the alarm off or, if you had a "rough night," you'd fling the alarm clock against the wall, hoping there was enough force that it would either a) shut the alarm off or b) smash it to smithereens.

The "Navy Years" (*Boot Camp*) *Reveille*

I don't know what Recruit Company Commanders (i.e., boot camp CCs) do today, but back when I was in boot camp (Orlando RTC, July 1973), the CC used a very effective alarm clock—a 50-gallon metal garbage can.

Hopefully, when he kicked it (more than once) down the middle of the barracks, it wouldn't bounce off someone's head. Most civilians have probably never experienced that type of wake up call, but it was very effective and got everyone out of their racks fairly quickly.

Of course, depending on the CC, you might also hear, "Reveille, Reveille! Drop your [fill in the blank here] and grab your socks."

The "Navy Years" (*Active Duty*) *Reveille*

Once you graduated boot camp, you were able to slack a little in "A" school, right? Well, welcome to *Reveille*.

I don't know about you, but it seemed like the Navy found the loudest man on the duty station to make an announcement over the 1MC, almost blowing out the amplifier tubes, to announce Reveille.

Yep, that was effective, too. Unless you figured out that you could put your pillow over your head to fall back asleep. Oh, wait…

NAS Agana Guam—Klaxon alarm

I was an ABH (Aviation Boatswain's Mate-Handler**)**. The ABH rate included aircraft firefighting. When I was stationed at NAS Agana, Guam ('73-75), I was assigned to the Crash & Rescue Division.

The "Crash Barn" was located right off the tarmac between the control tower and one of the squadrons (C-130, I think). We worked 24 hours on and 24 hours off—for 15 straight months, I might add.

Sure, we still had Reveille, but it was typically done over the 1MC by "one of us" in the Crash Barn. Sometimes, Reveille was loud, other times a little more "colorful," depending on who had duty on the phone desk.

But when there was an emergency, either an actual aircraft crash or an aircraft coming in with problems, we had a "Klaxon alarm" to sound the alert. For those of you who don't know what a "Klaxon" alarm sounds like, just watch a movie when a Navy submarine announces, "Dive, Dive." The "alarm" (ahh-oo-gaa) is a Klaxon alarm.

And before you ask, No, it wasn't quieter in the middle of the night. It actually seemed louder.

That-THAT-THAT was truly one way of waking people out of a deep sleep. Though I do remember Gary Borne, one of my good friends in the Navy (both Guam & *Kitty Hawk*), sleeping through one of those alarms one night while in Guam. We kidded Gary about that for a long time.

USS Kitty Hawk—Reveille

Once I transferred to *Kitty Hawk*, we were back to the old Reveille for wake up calls.

There were times, though, when you'd need an earlier wake up. That would be if you had an early morning watch, in which case you'd ask the guy who you were going to relieve to wake you or you had your own alarm. Back then, we used the wind-up alarm clocks.

Then again, sometimes there were those "special" alarms, coming over the ship's 1MC: "Man Overboard! Man Overboard! Port side" or "General Quarters, General Quarters, all hands man your battle stations!"

Yeah, those were fun times, too. Especially if you'd just hit your rack after an 18-hour shift on the flight deck.

The "Fire Department" Years—Klaxon alarm, part *deux*

When I was discharged from the Navy in '77, I knew I wanted to go back and finish college, but I still needed a job. Since I'd been a firefighter in the Navy, it made sense to join the local Gainesville Fire Department (Florida) when a job opportunity opened up, about a month after I got home.

Instead of 24 on and 24 off, like in Guam, we worked three shifts: 24 on and 48 off. That was great for me to attend the local community college full-time. Other firefighters also had another job or two.

Each of the seven GFD fire stations all had the old Klaxon alarm, too. I was already used to that wake up.

But one big difference from the Navy… at the #1 Fire Station, firefighters slept on the second floor. When the alarm went off after hours, we hopped out of our racks and jumped into our boots, which were right beside our racks. Just like back in Guam.

Back then, you didn't know which of the five firetrucks would roll until you got down the fire pole, then the Fire Chief would tell us which trucks were heading out.

Anyway, that was fun—20 of us, lined up ready to slide down one of the two fire poles. And this was before "pole dancing" became popular. But I digress…

That changed while I was still with GFD—the announcement came over the 1MC as to which trucks needed to respond. Then, you'd either hop onto the truck, hopefully before it started rolling, or go back to sleep.

I guess when I look back on it, the V-1 Division berthing compartment aboard *Kitty Hawk* was on the 03 level, directly below the flight deck, so I was one floor "below" the Crash fire trucks.

The "Working Stiff" Years *Reveille*

Once I graduated from college, I was a "working man," so to speak. Nothing like the Navy and nothing like the fire department, but definitely making a lot more money as an Electronics Design Engineer.

Now, instead of Reveille, I used an alarm clock. Not the bell ringer type I had during high school, but this new-fangled electronic marvel had a built-in radio to gently wake you. And it was so advanced, it had FM stations, too.

Since we only had one country music station in Gainesville at the time, it was set to that station. My wife wasn't so happy with "that" station selection to wake up to, which is why the alarm clock still sits on my side.

The "Kids Years" *Reveille*

Then came the kids. No need for an alarm clock now… we had the "kid alarm." Let me tell you, when you're sleeping in the middle of the night and your child wakes up crying, the three most wonderful words to hear from your spouse… "I'll get her." I stole that line years ago from somewhere, but it's still funny to me.

And just when you got used to having your baby "sleep in" past six in the morning, another one comes along, 2½ years later. The routine starts all over again.

I make fun of this, but we have two great kids. Well… they're not really kids anymore. Both are married and working professionals. I tell Becky all the time, "We must've done something right."

The "Retirement Years"—the *New Reveille*

We both turned 65 late last year and are now retired. "Livin' the dream," as we used to say in the Navy. We're happy in Florida, and staying healthy and keeping busy. That's what retirement should be, right?

Just when we thought we could start to sleep in, we had a new "alarm" show up on our front door a little more than a year ago.

Moosche is a 2-year-old stray cat that wandered up to our front door. We tried, but couldn't find the owner. No ID, no *nuthin'*. Despite attempts to re-home him, we/he finally decided we are his new family. He's a good cat. He really is…

… but he's an early riser. While I do have an alarm beside the bed (the same one above that we've had for 25+ years), and I do set it, I'm finding that it's really just a backup.

The "Moosche" alarm starts right around daybreak. When the sky is just starting to lighten and the birds start to wake up (love the sound of birds around the yard), we hear a slight "mew."

We "hit" the snooze button (i.e., we "ignore" him). Then a minute or so later, a little louder "mew." And if one of us is not up pronto ("Becky, I want to see two feet on the floor"), then we get the "Meoooowwww" alarm. That'll get you moving.

He is a good cat, *really* (I keep telling myself that). But he's not anything like my mom's gentle wake up. He's not Reveille, though I've tried to teach him to hum it. He's not a Klaxon alarm… yet.

He's just the newest type of alarm for us retired folks. And I'm getting used to it… and to him. I'm learning, slowly but surely, "People have dogs, cats have staff."

(22) The Go-Go, Slow-Go, and No-Go Years

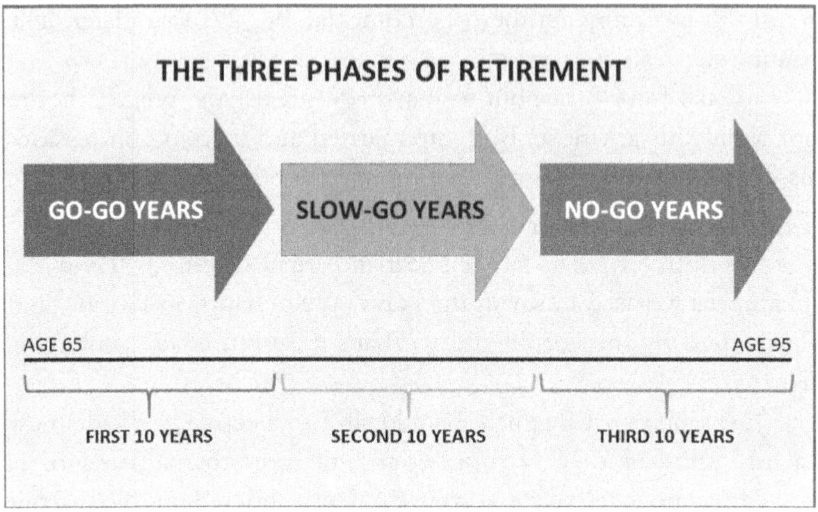

I'm writing this blog post on Friday, March 20, 2020.

When I started "A Veteran's Journey" blog last November (Veteran's Day), I had an initial list of about 30-40 blog "topics." That list is now up to about 60. I've found that when I write about one thing, something else may pop into my mind. And so, the list continues to grow.

I can think of all kinds of things to write about now regarding this current COVID-19 crisis. But that's already "well-covered" all over social media and the news, so I won't add to it. Instead…

… if I can continue to stick to a routine that helps me stay in the "normal" zone (whatever that is), then that's what I'm going to do. That, of course, may change in the not-too-distant future, but for now, here's this week's blog. I hope it brings you a smile and that it helps you through your day.

The year 2000 was *sooooo* far away

Back in the 70s when I was in the Navy ('73-77), I was still a "kid," like most of us were when we first joined the military. It doesn't matter if you were Navy, Army, Marine, Air Force, or Coast Guard. We were *all* young when we joined.

An 18-year-old kid with a lot of responsibility. That's how old I was when I reported to my first duty station, NAS Agana, Guam. I was an ABH (Aviation Boatswain's Mate-Handler) and assigned to the Crash & Rescue crew. I'd been through six weeks of training, had literally walked through fires in my training (yes, I was wearing an asbestos fire suit), and knew what to do and when to do it. Didn't we all?

Our job in Guam was to be on call if a plane crashed or had an emergency. We had five firetrucks and one tanker, if memory serves. Our motto was, and still is, "You light 'em, we fight 'em."

We were not only the firefighters for Navy aircraft, but we were also on call for civilian aircraft, since commercial jet liners used the same runways; the civilian airport was on the other side of the landing field. We probably had as many emergency calls for civilian aircraft as Navy aircraft; thankfully, no crashes while I was stationed in Guam.

In Guam, we ran two shifts: "Port" and "Starboard." I was Port. All that meant was that I worked the *first* 24-hour shift (yes, that was *supposed* to be a joke)... we worked 24 hours on and 24 hours off. And I might add, we did that for 15 straight months.

Initially, our shift change was at noon. However, not long after I arrived, the Division Chief (Bos'n Joye) changed it to 0800. That was a much better time to change shifts and gave us a *full* day for fun in the sun. After all, we were on a paradise island.

When I was 18 years old in 1973, I didn't think much about the year 2000. I don't think any of us did. It was "ages" away. I do remember talking with a good friend of mine, Glenn Law, about how old we'd be in the year 2000.

I figured since I was born in 1954, I'd be 45 years old. Holy Toledo, *that* (back then) was old.

Of course, now at the age of 65, my thinking has somewhat "shifted."

Glenn, BTW was with me both in Guam and aboard *Kitty Hawk*. He stayed in for a Navy career and (I believe) is living in the Oakland area. I was only in for four years—that's all I wanted to do at the time. Those were great years, though… many great memories.

Have You Ever Heard the Term…

Several years ago when my wife, Becky, and I were approaching retirement age, someone mentioned the term, "The Go-Go, the Slow-Go, and the No-Go" Years.

Neither of us had heard that before, but when you're planning for retirement, it's one of those things that you need to figure into your financial calculations.

From what I understand, the "Go-Go" Years typically occur when you first retire (for us, age 65). Some people do a lot of traveling, a lot of socializing, and a lot of visiting. That period of time usually lasts for about ten years (or that's what the financial planners tell you). Of course, it's very subjective, depending on health, finances, and your current situation.

After that, you hit the "Slow-Go" Years, where you still travel, but perhaps not as frequent and not as far. I get it, but I also know that many of us may not think in these "exact" terms. That's okay, just stay with me here.

The last "phase," if you want to call it that, is the "No-Go" Years. That's when you're typically in a retirement community, assisted living, or somewhere similar. You simply don't travel much, primarily because you don't move as well as you used to, or you need some assistance. It could be any number of situations. Again, this is all *very* subjective.

So, there you have it in a nutshell: the Go-Go, Slow-Go, and No-Go Years. At least, as it was explained to me.

But Wait, There's More!

Now that I've got you thinking…

I have a 91-year-old neighbor down the street who lives alone (his kids are out of state). He still travels (alone) to visit them, but I've also seen him working out in his yard, mowing, raking, pruning crape myrtles and hedges, and… on a ladder. I also see him at the local YMCA on the weight floor. He ain't in the "Slow-Go" or the "No-Go" Years yet.

Yeah, "be like Bill."

Becky's mother is 99 years old. She lives close to us in a retirement community—one that progresses from a house/villa/apartment to assisted living to skilled nursing. She and her husband moved into a house there years ago, knowing that was the plan: Go-Go, Slow-Go, and No-Go.

She's in the "No-Go" Years, but is still pretty peppy, still tries to write notes to family members, and relives a lot of old memories.

When she and her younger 90-year-old sister (who still travels alone from Tennessee) get together and start to reminisce, well… I just have to chuckle to myself when I hear some of the things these *young ladies* did when they were kids.

No, it wasn't the, "Here, hold my beer" kind of stuff, but for a couple of girls who grew up in the Boston area in the 20s & 30s, well, let's just say I was shocked more than once.

Yeah, "be like Marion and Doris."

"Unraveling" the *Later* Years

I've heard that getting old can be likened to unwinding a roll of toilet paper. This may not be an appropriate analogy given our current "TP" state of mind, but stay with me here. This is good.

When you get that fresh new roll of toilet paper and start to use it, it comes off the roll rather slowly. But as you get closer to the end of the roll, it appears to unwind rather quickly, wouldn't you say?

I'm not sure how that applies to getting older, but I think it's a relativity thing. But for me, it does seem that the older I get, the faster life seems to pass me by.

There's a great line in the movie, *Deep Blue Sea* (1999). "Preacher" (LL Cool J) says to "Tom Scoggins," (Michael Rapaport) a Harvard PhD: "Einstein's theory of relativity. Grab a hold of a hot pan, a second can seem like an hour. Put your hands on a hot woman, an hour can seem like a second. It's all relative."

Stay healthy, stay safe, and "wash your hands." We will get through this.

(23) Rough Seas: The Sea of Japan

I was a Third Class Petty Officer when I arrived aboard *Kitty Hawk* in March 1975. I believe there were two of us 3POs in Crash at the time: me and Gary Borne. Gary and I also served together at NAS Agana, Guam. Shelby "Big Willie" Williams was a Second Class Petty Officer, Jeff Atteberry was the First Class Petty Officer and Crash LPO (Leading Petty Officer), and CWO2 Tyrone Robuck was the Air Bos'n. Bos'n Robuck was the Crash HMFIC.

Westpac '75

During my one and only Navy cruise, Westpac '75, I was assigned to Crash. I would later become an aircraft director (Fly 3 Yellow Shirt) after we completed our year-long overhaul in the Bremerton Shipyards. But that's another story for another day.

The Philippines (PI, for short) was our overseas deployment homeport. This was *after* the Vietnam War ended, so as far as I know, we were never on *Yankee Station* (north) or *Dixie Station* (south), as they were called. But we did cruise around the South China Sea and the Sea of Japan. Different types of flight operations, but each with its own set of challenges.

We left PI in mid-October '75 and headed north toward Japan, starting our longest time at sea without hitting a port—about five weeks. Yes, I know some of you are laughing at that "longest time at sea" comment, and I admire those who experienced longer *at sea* periods, but this was Westpac '75 and the first *Kitty Hawk* cruise *after* the end of the Vietnam War.

Transit to Sea of Japan

It turns out that we were on the edge of a typhoon and we experienced high seas and a lot of rough weather on our way to Japan. All flight operations were cancelled. On the flight deck, we had to double all the tie-down chains on the planes. Instead of the usual six tie-down chains, each plane had twelve to fifteen. All the equipment, including the crash trucks, all the tow tractors, and even our Crash Crane, "Tilly," had tie-down chains.

During the day, it was so bad that if we ventured out onto the flight deck, we were to go in pairs with a life line tied between us.

We even saw water coming over the bow. I thought they were waves actually breaking over the bow, but the waves broke lower; it was just the force that pushed the water over the bow. That was impressive though, especially since the flight deck was sixty feet above the water line. I can't imagine how the smaller ships managed these horrendous sea conditions.

Yokosuka, Japan—Max Liberty

We arrived in Japan on Friday, October 17 and we were in for a treat; we would have six days of "max liberty" in Yokosuka. That meant that the only time we *had* to be on the ship (day or night) was if we pulled duty.

And since our Division had six sections, we only needed to be on board that one day; unless someone "volunteered" to take duty for someone else in the Division. That was pretty common, especially if the "duty fee" was good. Back in my day, that usually meant someone would pay $20 to take their duty for them for that day.

As we pulled into the harbor, it was a bright, clear day (the kind of day my mom would call a "Champagne Day") and we could see snow-capped Mt. Fuji off in the distance. It was truly breath-taking.

The exchange rate at that time was three hundred Yen for one US dollar. Japan was more expensive than PI and Hong Kong. A beer in Yokosuka cost 300 Yen ($1 US), compared to three pisos ($.45 US) in PI and $4 HK (Hong Kong dollars, about $.85 US).

Tokyo, Japan

On Sunday, Gary Borne, Bud Laney, Jim King, Tom Watson, Gary DeSauniers, and I took a day trip to Tokyo. We rode the bullet train and I'll tell you something, it *was* a bullet. Wow, I had never been on a train that fast before. It only took us thirty minutes. The cost for the train ride from Yokosuka to Tokyo was only 300 Yen ($1 US).

The "big problem" we faced was that it was Sunday and we had a hard time finding a place to exchange money. After an hour or so of wandering around Tokyo—and of course, *none* of us spoke Japanese—we finally found a small store that gladly exchanged dollars for Yen.

The weather was overcast and cold, but because we were walking everywhere, we kept warm. One of the first spots we stumbled upon was the *Emperor's Palace*, very old and very traditional Japanese. There was a huge moat surrounding the palace with lots of those humogonous gold fish.

We also spotted a Japanese McDonald's downtown. That was funny to us, since we were not used to seeing American businesses in foreign countries. We recognized the golden arches, but the rest of the writing was gibberish to us.

Japan was a welcomed stop. We had been in PI for a while and had visited Hong Kong, but Japan seemed more civilized, definitely cleaner, and had more of a business and cultural feel to it. When I look back on my time in PI, I realize I never traveled or toured anywhere while in the Philippines, which is really a shame because I'm sure there were some wonderful places to visit.

For some reason, I spent almost all my time either on the ship, the base, or Olongapo City. If I had to do it all over again, I would definitely find some other places to visit in the Philippines. That's what the lifers told me, too. You learn the ropes on your first cruise and you do more sightseeing and get more involved with the culture and the people on the second cruise.

Thieves Alley

Just outside the main gate of the Naval base at Yokosuka and across the street was "*Thieves Alley*." They say "Many a sailor has lost his credentials here," whatever that means. But there was a group of us, so I wasn't too worried.

By the way, does a sailor really ever have credentials?

Thieves Alley was loaded with bars and little gift shops and the owners liked to bargain. Years later, I'd visit Times Square in Manhattan and Bourbon Street in New Orleans, which would be a reminder of the bargaining power. Here, you could bargain away and usually come up with some pretty good deals. You had to be careful though, because while a watch *may* say Rolex, it may not really be a Rolex on the inside, if you know what I mean.

The bars were something else. I simply could not drink the beer, it tasted so bad. And it wasn't because I'd been drinking San Miguel beer in PI for so long, it just tasted like it had been "filtered through Luther's boots." But we did find a wonderful place that served great corn dogs and ice cream, something that the Crash Crew craved. We frequented that cozy little spot several times—no beer though, at least not for me.

Extreme Cold

Once we left Japan, we headed north through the Sea of Japan and circumnavigated the main island. It was extremely cold and it didn't help that we were flying around the clock. Since we did not have enough people in Crash to have two full 12-hour shifts, several of us pulled double duty, working twenty-four to thirty-six hour shifts. During Westpac '75, we had 14 in the Crash crew; we were required to have 11 on deck during flight ops.

We would catnap when we could, whether it was in the crash compartment on the vinyl floor or cramped inside the MB-5 crash truck. Back then, I think most of us could fall asleep almost anywhere and anytime.

In the South China Sea, the weather was always warm and all I wore on the flight deck was a pair of Army ODs, a red crash jersey, a red float coat, my boots and socks, and a cranial helmet.

It was quite a different story when we were cruising through the Sea of Japan in late October. I had my thermal long johns, my sweater, my Navy foul weather jacket, my rain gear and my aviator knit cap. Even with all that on, it was still freezing cold. We were working long shifts too and there were many nights when we could not get down to the mess hall for chow. We would get box lunches, but the old bologna sandwiches simply did not cut it.

Hot Soup & Cold Sandwiches

Several times during this trip out, I called down to the galley and asked if we could get some hot soup. They said sure, come on down. So a couple of us made several trips down several ladders to bring up a huge 25-gallon container of hot soup. It tasted so good and all of us got a break to come in and down some. It helped choke down the cold sandwiches, too. We always kept the coffee brewing, too. That is something we never ran out of.

There were many times when I would come in off the flight deck to warm up and to get a cup of hot coffee. When I started drinking coffee in Guam, I would have it with cream and sugar; there was *always* plenty.

But on the ship, I usually was the last one in to get coffee (I always let my crew come in first), there was never any cream or sugar, so I had to drink it black. I got used to that. I still, to this day, like my coffee hot, black, and strong. There were also many times when I would tip the 100-cup coffee maker to get the last dregs of coffee, complete with grounds. But in that miserable weather, even *that* coffee tasted good.

Launch the Alert 5!

We usually had several Russian "visitors" around us. Not only were there *Russian Bears* and *Badgers* flying overhead ("Launch the Alert 5!"), but we also had Russian trawlers cruising alongside *Kitty Hawk*. I had never seen one of those before, but they looked like ordinary fishing boats to me.

We waved, they waved. Hey, we're all friends here. This was during the Cold War and we did not trust them any more than they trusted us. There wasn't anything we could do other than carry on our regular duties.

Helo Down

Heading back to the Philippines, we lost another helo, this one to the water. I was on night shift, so I was in my rack sleeping soundly and heard the call over the 1MC, "Man overboard, port side."

At the time, I did not know what it was, so I quickly got dressed and mustered up on the flight deck in Crash. My friend Jack Kuiphoff was driving a tow tractor that day and later told me he drove over to the port side of the ship and looked down. The helo was on its side and the crew members were kicking out the windows and helping pull each other out—the helo was sinking fast.

In this case, apparently the helo was riding heavy and came in too fast and too low. Normally when a helo is going to land on the flight deck, the pilot brings the helo in off the port side *over* the water and *not* over the flight deck. The director then signals the pilot to fly sideways, bringing the helo slowly over the flight deck, ready to land.

They do this for this exact reason. Had the helo been coming in over the flight deck, it would have been quite a mess. They say the worst part of a helo crash is *not* the fire, but the spinning rotors that hit the deck and fly into a million pieces of deadly shrapnel.

The helo apparently experienced engine failure and was forced to ditch alongside the carrier. There were five men aboard—all were unharmed and were rescued by the plane guard helo. After everyone was accounted for by the "Man overboard" call, I was able to get back to my rack for another snooze.

USO—Double-wow!

The very next day *Kitty Hawk* received some surprise visitors. The USO had flown six beautiful women from the Miss America Pageant to *Kitty Hawk* for our entertainment. They had probably never landed on an aircraft carrier before, but they all looked great. I caught sight of them disembarking the COD (Carrier Onboard

Delivery) aircraft from atop the crash truck.

The only problem I had was that I had just come off the night shift, working through the night and into the morning and I was dead tired. While they were singing and dancing into the hearts of the *Kitty Hawk* crew, I was grabbing some much needed shut-eye. I did not get to see any of their performance which was done on a mock stage on the flight deck. But I later saw some great photos and heard it was a wonderful show.

Japan was one of those treasures I remember visiting. While I haven't made it back yet, it's on my list. There are so many beautiful places in this world to visit, as well as the good ol' USA. I'll get there… Just not today.

(24) Hurry Up & Wait!

Holy Toledo! Have veterans been training for this current COVID-19 crisis all this time? I know all of my Three Years, Eleven Months, and 29 Days—But Who's Counting were well spent helping me prepare. Even though that was so long ago (1973-77), I still remember. Don't you?

This is usually one of the first phrases a new recruit learns early on after joining the military. Navy, Army, USAF, USMC, or Coast Guard—it didn't matter; we were all in it together.

Back then, it meant that you were always told to "hurry," but when you got to where you were going or ready to start a task, you then had to "wait." And like anything else in the service, "it" rolls downhill.

A Prime Example

Here's an example: "Cruise photographs are to be taken of all

5,500 sailors, starting at 1000." This was a "scheduled" photo session, meaning that different divisions would have their photos taken at specific times during the day. The *Captain* wants every sailor to be dressed in summer whites (it was Westpac '75). V-1 Division was part of the Air Department and we were slated to go first.

We were to muster for the cruise photographs at 0930. The *Division Officer*, anticipating that there may be a "few" sailors in our division that will slow things up, instructed everyone to muster at 0900. The *Division Chief Petty Officer*, knowing *exactly* what the Division Officer *really* wanted, had everyone dressed to the gills and standing at attention when the Division Officer arrived for inspection. So he instructed all sailors in V-1 to muster at 0800.

And that's how the phrase, "Hurry Up & Wait" came about.

"Live & learn," they said.

"Living the dream," I said.

Of course, this is *not* just the Navy, but *every* branch of service. As a general rule of thumb (I'm sure you've heard this before), if you're 15 minutes early, you're on time. If you're on time, you're late. And if you're late, you're screwed. Face it, it's the truth and we've all be there before.

Duty Stations

Once I got into the "real Navy," that is, a duty station (for me, NAS Agana, Guam and later *USS Kitty Hawk*), it became almost a daily mantra. Here's just a few examples of my experiences with "Hurry Up and Wait." What about yours?

Chow line. *Kitty Hawk* had two mess decks: forward (fast food) and aft (not so fast food). There was also the Wardroom for officers, but I was enlisted and never set foot in one. When *Kitty Hawk* was at sea, there was usually a long (LONG) line at the chow hall. Face it, with 5,500 men to feed, there's a wait—suck it up, Buttercup.

Every now and then, flight deck personnel got a special "mess chit" to go to the head of the line, especially during long flight ops. All that meant was that we were able to change from standing in a long (LONG) line to standing in a long (SHORT) line. A big, burly Master-at-Arms would be there to make sure you had a legit mess chit. I don't know about you, but I read a lot while standing in line waiting for chow. I think I probably read a book a day; well, maybe a book every two days.

Sick bay. Lines weren't as long for Sick bay as the chow lines, but sometimes they didn't move as quickly. It didn't take me long to figure out that the lines for Sick bay were usually longer *AFTER* a port call than before. I guess it's because of the various illnesses one may "pick up" while in port. Actually, there were two *common* symptoms *AFTER* a port call: nausea from too much drinking and that "other thing" from too much…, well, you know…

Liberty call. I finally figured out why the After brow was so narrow (one sailor going, one sailor coming). When you've been at sea for several weeks and have the "itch" to get off the ship and explore the local "culture," there's a mad rush to disembark.

Of course, there's always the "Salute the Flag," and "Permission to go ashore, sir!" traditions, but with 5,000 of your closest friends all waiting in line to go ashore for some fine food and dancing, it's Hurry Up & Wait time.

What about when you're *not* aboard

Hurry Up & Wait takes on a whole different meaning when you're not aboard.

Movies. Right before I was discharged in July 1977, a new science fiction movie came out: *Star Wars*. It's now called, *Star Wars: Episode IV—A New Hope*. For me, it will always be the *best Star Wars* movie in the series. I was still aboard *Kitty Hawk*, a "short timer" with less than 30 days before my discharge.

When *Star Wars* came to one of the San Diego theaters, several of us climbed into a car to go see it. The line was literally wrapped around the movie theater, which was located in a mall, so that meant—in terms of stop lights, if you follow me—at least 2 lights before turning left. In other words, we'd be waiting for the next showing or two, just to buy tickets. Now, there's this fancy, schmancy thing called *Fandango*, but back then… nada. You stood in line to buy tickets.

Restaurants. Over the years, I've wanted to try several new restaurants that I've known were popular. "Hurry up, let's get there before there's a long line," only to get there and find out there's a long line.

My wait limit *used* to be about an hour. As I've gotten older… "It'll be about 45 minutes, sir." "Fine, put our name on the list and we'll go somewhere else" (you know… just in case you don't find anywhere else). Don't tell me I'm the only one on this planet that does that. My wait limit has definitely decreased as I've grown older.

Concerts. There was that time when Linda Ronstadt and the Eagles were appearing here in Gainesville at the University of Florida football field in the late 70s, right after I got home from the Navy. Yeah, I waited in line for what seemed like hours to get tickets. It wasn't a "nut to butt" line like in boot camp, but it was pretty well packed. And yes, it was well worth the wait.

Airports. "We need to leave now to get to the airport early enough to park, walk to the terminal, get our boarding passes, check our baggage, get through the security checkpoint, and get to our gate, only to find out the flight is delayed… what, an hour?" Crap.

Well, at least we showed up on the day we were supposed to fly out, as opposed to that time years ago in my "consulting days," when I was traveling so much and was so "out of it" I went to the airport *a day before* my 6 a.m. flight. That was an interesting story, especially when I went back home and my wife asked, "What's up?" Don't tell me that I'm the only one on this planet that's done that before.

DMV. Don't get me started with the Department of Motor Vehicles. I'm sure I'm not the only one who's had to make more than one trip for a driver's license renewal.

And my wife… holy schmoly. "You mean your passport doesn't have your maiden name? Didn't you bring your marriage certificate to verify you've been married to the same man for 39 years? And when you come back, make sure you know where you left off…" you know, like "Hurry Up & Wait."

Our Current Situation

And now, with this COVID-19 pandemic, I've learned that the Hurry Up & Wait training I had in the military has helped me better cope with this "unknown/unseen" reality. If nothing else, it helps me remember some of the more humorous times in the service.

Grocery stores are now open early for "senior citizens." For me, that just means the line is longer and moves slower. But at least there's a better chance for getting my weekly roll of toilet paper before the shelves are empty.

I thought I was being smart when I set the alarm for 6a in order to get to the store at 6:30a to get in line before they open at 7a. When I got there, the line was wrapped around the corner.

Of course, with those little blue painter's tape crossed taped on the sidewalk to keep that six foot "social distancing," it's a reminder of what a different "normal" we are in. And yes, I've learned to go at different times of the day to different stores and we've gotten to where we only go to the grocery store once a week.

Doesn't this remind you of the lines at the gedunk shop aboard the ship? Yeah, you remember *those* lines, trying to buy a carton of cigarettes for $10 or trying to buy a can of sardines and Cheez Whiz because you missed chow.

Restaurants, take out only. I don't know about you, but in the past, when we went out for dinner, we usually frequented the same 5-6 local restaurants, since they are our favorites. Over the years, we've figured out what time to go and how long the wait will most likely be (see my "wait time" tolerance above). And yes, as we get older we tend to go earlier. What about you?

Now, there's only take-out available. I get it and I certainly understand and I don't want to put anybody, especially me & Becky, at risk. So, take-out is the way *to go* (pun intended) right now. But it's thrown off my "timing" of when to go. Do we go earlier to beat the crowds, or do we go later? It's a whole 'nother world out here right now.

Liquor stores. Currently, liquor stores are open here in Florida (yes, they ARE "essential services"). I could go on about the essentialness of a good tequila, scotch, or wine, but I digress. No longer is it a problem for me to go early & go often (my current motto).

Get there early to beat the crowds, in and out. I know what I want and where it's located. My current "record" is about 60 seconds. That is, until I get to the counter and find another "older than me" gentleman in line ahead of me. "Maintain that six-foot social distancing." Yes, I know, but I can only hold my breath for so long here.

The "New Normal"?

In today's "new normal" during this COVID-19 crisis, I don't hear "Hurry Up & Wait" too often. However, I certainly feel it day in and day out. When I'm standing on the blue tape "X" six feet behind the next person in line waiting to go into the store to buy a dozen eggs or a loaf of bread, if they have them, well… I don't say it out loud. But I sure think it. Am I alone here?

It's almost like that training we all began years ago is now starting to pay off. "Hurry up and wait." Yeah, I've been there and done that. It's a little different now than back in the 70s.

None of us know how long this will last. I certainly understand the uniqueness of this virus and therefore the unknown dangers. And yes, I'm in that category of… Lord, I hate to say this… age 65+ "higher risk" (my wife & I just turned 65 late last year).

And even though we don't feel "old" (whatever that is), I am trying to stay aware of the situation, stay home and only venture out for groceries when I have to. We do get outside for long walks, usually one or two a day, and that helps clear the mind, clear the soul, and get the day going. It's a routine for us and we are fortunate enough to live in a neighborhood with lots of good walking space.

It's not the same being on a flight deck, watching the sun rise or sun set, but for now, it'll do.

A couple of weeks ago I wrote about the "Go-Go, Slow-Go, and No-Go Years." That is, what some people call the retirement phases. Well, I can tell you that it's now been reversed: we're currently in our "No-Go" *Months*, and when this thing lets up, we may get to our "Slow-Go" Months, then we'll get to the "Go-Go" *Years*. At least, that's the current plan, subject to change, of course.

"Hurry Up & Wait"—yep, been there and done that and have a T-shirt. It looks like we're doing it again.

(25) "Livin' the Dream"
Life Aboard an Aircraft Carrier

I was going to blog this week about a liberation that happened during WWII…

April 14, 1945—75 years ago today, my father was one of many who helped liberate the Buchenwald Concentration Camp, near Weimar, Germany.

It *was* my intent to blog in detail about that liberation, of what he and others did, what they saw, and what they came away with. I wanted to honor the Allies and the men of the 80th Infantry Division. In fact, I had already written the blog and was ready to post.

However, because of everything we are all facing during this current pandemic crisis, I'm going to hold off posting about Buchenwald at this time. Even though these events happened 75 years ago, it's something that we should Never Forget. Instead, I'll just post a photo I took of the main gate when Becky & I were there a few years ago.

And now, to this week's blog post…

Aircraft Carrier Flight Deck Operations

I was watching *Top Gun* last night… well, to tell you the truth, I really only watch the first 15 minutes, since that part involves fighter aircraft, launches & recoveries off a carrier, and other awesome random flight shots. After that, well… let's just say I've seen it. The *best* part of the movie is the first 15 minutes.

In fact, my wife now knows *everything* there is to know about flight deck operations by listening to me explain every little minutiae, every time *Top Gun* comes on.

It finally got to the point where I quit pushing "pause" after a short 3-second clip and spending the next 5 minutes telling her about the holdback bar, the shuttlecock, the "Shooter" signals, and why there are so many men with different colored jerseys on the flight deck.

After a dozen or so times of glancing over to her "rolling eyes," (you know *that* look) I figured the first 15 minutes of the movie was enough.

Am I the only airdale that does that?

There are plenty of movies and documentaries about flight ops aboard an aircraft carrier. But what else does a sailor do aboard the ship at sea, when *not* on duty? In my case, when outside of flight operations. I know it varies among all of us, but here are a few things I remember most during my personal "down time."

Bow Flying

Even though I spent most of my working days on the flight deck of *Kitty Hawk*, I also spent a lot of my "off time" on the flight deck. One of the things I remember vividly, and I'm sure most everyone did the same at one time or another, was to stand on the bow and lean forward over the safety nets. Back then, we called this "Bow flying."

I'm sure there are several of you that "accidentally" fell into the safety nets, especially at night, but I never had the nerve. There's something about having that fresh, clean, salty air in your face. And if you were wearing a foul-weather jacket, you could unzip it, hold the bottom out like wings, and almost "fly" over the water.

Even today, when I'm walking outside and a stiff wind blows in my face, I close my eyes and… *I remember.*

When my wife & I went on a cruise a few years ago (my first, since leaving *Kitty Hawk*), I tried to demonstrate the technique on the bow of the cruise ship. It just wasn't the same, since there was a 4-foot rail in the way. But it did bring back some great memories.

When my kids were younger, I tried to explain the "flying" sensation, but they just rolled their eyes. However, when the movie *Titanic* 1997) came out and the scene—you remember *that* scene—when *Jack* (Leonardo DiCaprio) and *Rose* (Kate Winslet) were on the bow, hanging out over the railing… flying… My kids finally understood what I'd been trying to explain to them for years.

Poker?

A lot of my buddies played poker during their down time. I was never any good at it. I'd play Hearts or Spades, but not poker. I know some guys would come off the flight deck, play almost all night, get a couple of hours sleep, then be back on the deck for another day of flight ops.

My tall, lanky friend, "Stretch," a fellow Yellow Shirt director, told me one time he played poker all night and lost $2,000. Then the next night, he'd win it all back. And hence the term, "It's not just a job, it's an adventure."

The Smoking Lamp is Lit!

I read somewhere that on a good day, you can see about seven miles to the horizon at *sea level.* When you're on the flight deck of an aircraft carrier, about 60 feet above the waterline, you can see about twelve miles.

At night, we used red lights in the compartments that opened up to the outside of the ship, thus reducing the amount of ambient light. We didn't want to alert any "non-friendlies" who may be shadowing the ship.

I also read that at night, you can spot a lit cigarette at two miles, but I don't know that for a fact… I didn't smoke.

When you heard the announcement, "The smoking lamp is lit," it was like a 1MC alarm for people to scatter about the ship to the various *designated* smoking areas. Yeah, right. I know guys smoked in the catwalk—there were those "designated" hidden spots.

Fortunately for me, I never picked up the habit. I'm thankful for that, especially since both of my parents died way too young of smoking-related diseases.

The Forecastle

I know most people want to pronounce this, "Four castle," but for the life of me, I can't… I just can't. It's the "Fokes hole," dadgummit.

And it is one of the most interesting places aboard the ship. I spent a little time in the forecastle, peering out through the portholes where the anchor chains pass. That was quite a sight, watching the bow cut through the water like a knife through warm butter.

I didn't realize it until I researched for my Navy book that each link of chain on *Kitty Hawk* weighed about 360 pounds and there was enough room for more than a thousand feet of anchor chain. That's a lotta chain.

The Bomb Farm

On the starboard side of the island, just outside of the Crash compartment, there's an area designated as the "bomb farm." This is where the aircraft squadron ordnance crews stored their armament.

It took a little while for me to get used to walking out the Crash compartment and seeing dozens of these 700-pound bombs side by side, next to Sparrow and Tomahawk missiles.

The Sparrow missiles were about ten feet long and about six inches in diameter. The nose cone was made of ceramic material that could take the heat of friction during supersonic flight.

They were also great for sharpening knives and since I always carried a four-inch Buck folding knife with me, I would stop every now and then and sharpen my knife on the missile nose cone. Of course, no one was around to tell me not to.

Before I got to *Kitty Hawk*, one of the Crash crew decided he wanted a "personal" knife sharpener. The rumor was that he walked outside the crash compartment one night and whacked the nose cone of a Sparrow missile with a ball-peen hammer, cracking it into several large pieces. He never got caught for it, but when he left, he forgot to take his "souvenir" with him.

I was assigned his locker when I first checked into Crash, so I was the proud owner of a piece of a Sparrow missile. I still use it as a knife sharpener, too. The curvature of the piece of cone fits right on my thigh—perfect for sharpening knives.

Sunrises, sunsets

Perhaps the most special memory I have of being at sea (*not* including flight operations) is the beauty of watching a sun rise or sun set from the flight deck. When you're at sea in clear blue ocean water, it's easy to lose yourself in the beauty of God's world. On clear, calm days, the cool breeze coming over the bow, the slight smell of salt in the air, and watching the sun in the background… I just get lost trying to describe it.

There are tons of photos on many of my Facebook Navy groups and they're all better than mine, but nothing will replace the feeling of being there, on the flight deck of *Kitty Hawk*, taking it all in. Those are probably my best memories that I can easily "pull back" from years ago.

(26) That Adrenaline Rush

Can you remember way back… back *before* high school graduation when you experienced an "adrenaline rush?" You know… Cars, competitive sports, hunting, fishing… Or, maybe that's the "guys" perspective. Stay with me here…

Most of *my* adrenaline rushes in those early teen years came from driving my mom's 1966 Oldsmobile Cutlass SS, or my sister's boyfriend's '69 GTO. Not that I drove *too* fast mind you, though I did manage to rack up four traffic tickets *before* graduating high school. Of course, *none* of them were *my* fault.

I didn't get too many "rushes" driving my dad's old 1954 Willys Jeep at a top speed of 45mph. Unless you count the time when it slid down a muddy road and flipped over.

Thank God, the Jeep *had* a hard top. After 13 stitches in my head and a couple of days on crutches, I got smart and installed a roll-bar—or maybe that was my mom's idea.

Yeah… *those* were the days. And I couldn't even spell "adrenaline" back then. But I digress…

Then Came the Navy…

I was only 17 when I started college at Florida State University in the Fall of '72. I flunked out after my first year—"dismissed for academic reasons."

I was trying to figure out what I wanted to do with the rest of my life—you know, those decisions you make when you're 18 years old. That's when I joined the Navy. I knew that I'd have a good time, though: "It's not just a job, it's an adventure."

Boot camp (Orlando RTC, Company #163, '73) was "interesting." That's where they break you down as individuals and build you back up into a cohesive team. "A" school was more interesting (Aviation Boatswains Mate-Handler; Lakehurst, NJ, Fall '73), where I trained for my Navy "job": aircraft directing and aircraft firefighting.

When I got to my first duty station, NAS Agana, Guam ('73-75) and assigned to Crash & Rescue… well, then it got *really* interesting. That's where the "real" Navy began for me.

Crash had to maintain a presence on the airfield whenever aircraft were flying, so we spent two, 4-hour shifts in a 24-hour duty day in what we called the "Hot Spot." Of course, the *adrenaline rush* didn't come from sitting in a Crash truck on the Hot Spot between airport runways watching aircraft take off and land. When you're a Crash Firefighter, the adrenaline rush comes when something "not-so-good" happens.

I don't know about you, but continuous training was a *big* part of my Navy career (all four-years-minus-one-day). Training for those things, you know… things that you hoped/prayed wouldn't happen. But that's why you trained… just in case they did.

Aircraft Emergencies…

At NAS Agana, Guam, whenever we got a call for an aircraft emergency, we never knew if it was an aircraft that had called the control tower with a problem, or if it had already crashed. As soon as the Crash alarm went off and we climbed onto the fire trucks, we could usually tell—if there was smoke, there was usually fire. If no smoke, then it was a different kind of emergency, like a fuel spill, a blown tire, or an aircraft coming in with an engine shut down.

Did you know that the old Boeing 707 aircraft had four engines and that it could fly on just one engine? Yeah, me neither. But fortunately for them (and for us), one like that landed in Guam with minimal fanfare—other than five Crash firetrucks chasing it down the runway after it landed until it reached the civilian airport terminal on the other side of the field.

But *that*, for me, was an adrenaline rush.

An "Oh Crap!" Moment…

Another one—at NAS Agana, Guam, we got an emergency call for a commercial 747 jet liner coming in, fully loaded with more than 400 passengers.

It was a dark & stormy night (no kidding, it was about 9pm, and it was raining cats & dogs). We weren't quite sure what the emergency was at the time.

We were all lined up on the taxiways (NAS Agana, Guam shared the same *parallel* runways with the Guam civilian airport). Each Crash truck was assigned to wait on a specific taxiway for the flight to land. When it touched down, the first truck would take off, chasing the aircraft down the runway, then the next truck would follow after the aircraft passed its taxiway, and so on.

Normally, we could tell how far out the aircraft was by their landing lights. However, because it was raining so hard, we couldn't see this aircraft approach until it was right over the end of the runway.

We immediately noticed it was way too far to the starboard (right) side of the runway—that's the side where we were parked & anxiously waiting. The Crash trucks were literally parked about 30 feet away from the runway.

When that 747 landed, its starboard landing gear touched down on the lights that lined the runway. Its starboard wing would pass inches away from the top of our Crash truck roof.

Like a synchronized swim team, *every* Crash truck driver (including me) slammed their trucks into reverse and put the pedal to the metal, all quickly backing up. The fire chief was in a pickup truck and I saw his tires smoking, he'd hit the gas full throttle.

Fortunately for the flight, and for us, the aircraft landed safely, though it took out about half the runway lights. That was what I would call an *adrenaline rush.*

The Flight Deck: Adrenaline "Levels"

Working on the flight deck of an aircraft carrier was one of the most exciting times in my life. I've written about many different events and times aboard a Navy ship.

I've sometimes describe the flight deck as a constant source of adrenaline, some events raising the "levels" of adrenaline.

Aircraft launches and recoveries were always exciting and were what I'd call a "Level 1" adrenaline rush—just being there. An ABH has two main jobs: Crash & Salvage and aircraft director *Yellow Shirt.* Both were equally exciting to me.

Level 3 Adrenaline Rush

During the Westpac '75 cruise, an A-7 Corsair launched off the bow catapult, but it left a 700-pound bomb behind on the flight deck. I'd call that a "Level 3" adrenaline rush.

As a firefighter, you wait to "go to work" when something bad happens. This was "bad." In this case, we were parked mid-ships and we all saw the bomb slide to a halt. I was driving the Crash truck that day—truly an "oh crap" moment.

Thank God, the bomb did *not* explode. Long story short, the EOD (Explosive Ordnance Disposal) team came onto the flight deck and gave it a look-see and let everyone know it wasn't armed. That dropped the adrenaline level down a notch.

Level 4

There was that time when I was on the "angle" as a Yellow Shirt director when an A-7 Corsair grabbed the arresting cable (normal recovery). Not realizing he had, the pilot applied full power to take off again. He was about ten feet in the air when suddenly the A-7 slammed back onto the flight deck.

Thank God the arresting gear cable didn't snap. The aircraft was fine, but I think the pilot went in for a "skivvy check." That was what I would call an adrenaline "Level 4."

I've also seen an F-4 Phantom launch and for some reason its elevator froze and the aircraft went vertical, about 100 feet off the water.

Even with full afterburners applied, it wasn't going anywhere. You could tell the pilot was struggling to maintain control, but he simply couldn't. I watched the pilot and the RIO (Rear Intercept Officer) punch out (eject) and land safely in the water to be picked up by the rescue helo.

Level 5

What's an adrenaline "level 5" (the *top* level)? I've *never* experienced that. I've seen plenty of training videos, but I never (knock on wood) actually experienced an aircraft crash upon landing. And I'm not sure I'd want to, though I know my training would kick in if it happened.

Note/disclaimer/don't sue me: These Adrenaline "Levels" mentioned in this post are my own definitions. They're not scientific by any means and are my weak attempt to share with you the various levels of adrenaline I've experienced in my careers. I'm quite certain you've got your own stories; feel free to share them with us.

Your Training Kicks In…

In the Navy (and later in my time as a Gainesville City Firefighter), we constantly trained in order to be best prepared for *any* event/emergency that happened. There's no way that you can train for any and all possible events.

But with enough training and with a good crew and a dynamic leader, your training automatically kicks in when needed to handle almost any situation. I truly believe that confidence stays with you throughout the rest of your life. "One step followed by another…"

One thing I remember about all that training is the amount of *awareness* you naturally develop over time. I'm not just talking about awareness for your job; I'm talking about your constant overall awareness of your surroundings. On the flight deck, it was imperative that you "keep your head on a swivel." That awareness seems to have stuck with me through the years. How about you?

Today, I find myself being even more "aware" of things today I didn't really worry about before a few months ago. This COVID-

19 crisis has changed everything and I'm learning to adapt to the "new normal." Grudgingly… but I'm adapting.

When we go outside for a walk or a bike ride…. Sunglasses-*check*; *USS Kitty Hawk* ball cap-*check*; face mask & bandanna (in case I need it)-*check*; hand sanitizer (in case I need it)-*check*.

I'm also more aware of people around me. When we pass someone on the sidewalk or standing in line at the grocery store, my head is constantly on a swivel, just like on the flight deck. And that constant reminder, "Use the hand sanitizer *after* grocery shopping but *before* opening the door to your car." The new list can become endless.

The "New Normal" of Awareness…

This is a strange "new normal" we're in these days. With my Navy training and my confidence, I sometimes feel my awareness is in overdrive. Good thing we're homebodies. It comes in handy.

And oh yeah… before I forget… one more.

I remember there were times aboard *Kitty Hawk* where the head was low on non-skid. I quickly learned to keep an extra roll in my locker… you know… just in case. Who would think that 45 years later, I would become aware of how many rolls of toilet paper we use in a week… then calculate how many weeks we had left before running out? Am I the only one who does that?

Yep, all those years of training & awareness still come in handy.

(27) Time-honored Navy Traditions

I'm a sucker for traditions, especially military traditions. While I only spent four years in the Navy, I was fortunate to have experienced many time-honored traditions.

"Traditional" Military Traditions

Some Hollywood movies actually do a decent job of showing "traditional" military traditions; other movies, meh. In the Navy, these movies may introduce you to a Burial at Sea or Manning the Rails. But there are many (MANY) other traditions that you may not know about.

Reveille, Retreat, and Taps

Most everyone is familiar with the term "Reveille." It signals the beginning of the day, usually 0700. I wrote a blog earlier about "Reveille and Other Alarm Clocks Over the Years."

"Retreat" signals the end of the *work* day, 1700 hours. In civilian terms, 5:00 p.m. During Reveille and Retreat, if you're outside, military personnel stop and turn in the direction of the flag. If in uniform, they salute the flag. *Reveille* and *Retreat* are methods of paying respect to the US flag, our country, and to our fellow military personnel, past and present.

"Taps" is a signal for the end of the day, at 2100 or 2200 hours (9:00 or 10:00 p.m.), depending on the branch of service. *Taps* is also a signal to begin "quiet" hours. It's a bugle call, meant to *honor* all military service personnel who paid the ultimate price.

Boarding & Disembarking a Navy Ship

When a sailor boards a ship, she or he first faces the stern and salutes the flag. Then s/he turns to the officer on watch at the brow, shows his military ID card, salutes the officer and asks, "Permission to come aboard, sir?" The officer on watch smartly returns the salute with, "Permission granted."

When disembarking a ship, the reverse occurs. The sailor shows his military ID, salutes the officer and asks, "Permission to go ashore, sir?" Again, the officer on watch returns the salute with, "Permission granted." The sailor then salutes the flag and walks off the brow.

Both officers (Forward Brow) and enlisted personnel (After Brow) follow this tradition. As a Petty Officer, I've stood duty on the After Brow and watched many sailors, drunk as a skunk, stagger up the After Brow and still manage to salute the flag and the officer on watch.

Burial at sea

A couple of weeks before *Kitty Hawk* left for Westpac '75, I had the opportunity to witness one the Navy's oldest traditions: a Burial at Sea. On April 24, 1975, the late Lieutenant J. J. King was committed to the sea from *Kitty Hawk*.

Those attending the ceremony wore their dress whites, initially standing at attention, then at parade rest. A burial service at sea typically includes a few prayers, a reading of scripture, the committal, and the benediction. Then the Marine guard fires three volleys, followed by a lone bugler playing Taps.

I watched this solemn and momentous ceremony from *Vulture's Row*, an outside observation deck on the port side of the ship's island.

Manning the Rails

When a ship comes back to home port from an overseas tour, sailors "*Man the Rails*," another formal Navy tradition. Everyone who participates dresses in the Uniform of the Day. Sailors line up arm's length, called "Dress Right," around the edge of the flight deck, facing out.

If the ship was coming into port, we'd stand at parade rest unless there was a special memorial, such as passing the Arizona Memorial in Pearl Harbor. Then, we would come to attention at the appropriate time. Honoring those sailors aboard *USS Arizona* who gave their lives was a memorable event which I'll never forget.

Returning from Westpac

When *Kitty Hawk* returned to San Diego from Westpac '75, we manned the rails. I felt a sense of pride and was happy to be coming back to the states after being away for six months. In this case, since we returned in December, the Uniform of the Day was winter blues.

Approaching the pier at North Island, we saw hundreds of Navy wives and children waving with signs of "Welcome Home, We Love You." The signalmen on deck were signaling to their wives, who, in turn, signaled back. We didn't have cell phones back then, so everyone with anyone was desperately looking, trying to spot a familiar face in the crowd. Then you could see the joy in their eyes and the frantic waving, "I'm here, I'm here."

The new fathers got off the ship first, greeting the moms and their babies with open arms and tears—lots of joyful tears. Some of the older guys who stayed on the ship teared up. Even though I was still only 20 years old without a wife or girlfriend, it was an emotional and touching scene.

A few wives had flown to the Philippines, Hong Kong, or Japan for a brief visit with their husbands. But this was the first time most had seen each other in six months. It was truly wonderful to see so much love and affection in one place.

Pollywogs & Shellbacks

During Westpac '75, more than once we heard scuttlebutt that we were heading south to Christchurch, New Zealand. But alas, that never happened. Since Westpac '75 was my one and only Navy cruise, I never had the opportunity to cross the Equator. Hence, I'm still a Pollywog.

If you haven't heard that term before, stay with me here. It's going to get interesting.

Whenever a Navy ship crosses a certain point on the globe, all sailors who've never done so are "initiated" into an unofficial fraternity of sailors who've been there before. The most famous example of these ceremonies is the moment a ship crosses the Equator at any point in the world. Those who have crossed the Equator and celebrated this ceremony are called "Shellbacks."

The Court of King Neptune

Officially, it's called "Crossing the Line," but sailors know it as "The Court of Neptune." The uninitiated (known as "Polliwogs") must bow before King Neptune (usually the ship's captain) and entertain his queen, Davy Jones, the Royal Baby, and his dignitaries.

The older and more experienced salts, at least those who were Shellbacks, took over the ship, including the officers. If the CO had not previously crossed the equator, he was a Pollywog too, and would take part in the ceremony.

It would start early in the morning. Even though the Pollywogs expected something to happen, they didn't know when or where. The Shellbacks in each division would take care of their own, getting them up out of their racks by various means, including using shaving cream, talc powder, eggs, salt water, or any combination of the above.

Shellbacks would dress in different types of costumes and put on face paint, each to his own. Many dressed to look like pirates. Pollywogs would usually turn their uniforms inside out.

They were taken to either the hangar deck or the flight deck (if on an aircraft carrier), depending upon the conditions of the ship's mission. There they would be subjected as a group to other various harassments, including swimming on the flight deck on their bellies, constantly being sprayed with salt water from a fire hose, or worse.

There would often be a ritual of crawling through a sewage

tube, made from old rotting vegetable material the galley had been brewing for a few days, and crawling up to the *Royal Baby*—the biggest, fattest, hairiest, and ugliest sailor on the ship. There, you would have your face rubbed in his belly full of axle grease or something equally disgusting. I know the activities varied from ship to ship, but you get the idea.

All of this was in good fun, of course. When you completed the ritual, you received a certificate, signifying you had become a *Shellback*. This is something that sailors cherish their entire lives.

You think you've gone through hazing when joining a fraternity or a sorority? This is hazing like you have never seen nor heard. Note that the Shellback ceremony *may* have changed since my time in the 70s.

Lesser Known "Traditions"

Balloon Shaving Cream—Sometimes we just wanted to have fun with a new recruit, especially if *said recruit* copped a pompous attitude. One of the more painless methods of humiliation was what we called, the "balloon shaving cream" trick.

Several of us would gather around the recruit's rack in our berthing compartment. His privacy curtains would already be down. We would fill up a balloon with shaving cream, blow it up a little, then while guys held the curtains down, one of us would open the corner flap and let the balloon loose in his rack.

Once the balloon finished its "duty," we would all split and run in different directions—he didn't know what hit him. I have to admit, that was one of the funnier things we did on the ship.

Launching the Shooter's Boots—After our one-year retrofit in the Bremerton Shipyards (1976), we returned to San Diego. One of the catapult officers ("Shooters") retired from the Navy.

During a "family cruise day," he brought his ten-year old son on board, took him around and introduced him to everyone. He was a cool little kid—very confident and cocky, just like his dad.

One of the traditions on an aircraft carrier is that when the Shooter retires, he launches his flight deck boots off the catapult.

It was heartwarming to watch the Shooter and his son go

through the routine of a standard cat launch, including the final salute to the boots, the hands in the air, swinging fingers, and then the launch of the boots. His son had such a great time mimicking his dad and it brought back a lot of memories of me and my dad when I was that age.

Changing Duty Stations—It was a tradition to "hose down" a short timer on his last day of duty before he transferred to another duty station. That meant a bunch of us would wrestle him down out in front of the Crash Barn (NAS Agana, Guam) and hose him down.

Sometimes we'd use a garden hose, but most of the time we tried to get him with one of the crash truck hoses. I found it quite funny since this would probably be the last time we ever saw these guys.

Promotion to Petty Officer—It is a big deal when a sailor receives a promotion from a Seaman/Airman (E-3) to a Petty Officer (E-4). We always partied when one of the guys in our division got promoted. As a matter of fact, we would always party even if nobody got promoted. After all, it is a Navy tradition to drink, plain and simple.

Sometimes the promotee would drink a little too much. The drunker he got, the easier it was to "handle him." By that I mean there were several times when a newly minted petty officer woke up at the Crash Barn (NAS Agana, Guam), tied tightly to the telephone pole, and greased up from head to toe with standard Navy issue axle grease. That stuff is hard to get off and takes several weeks to get the stink out.

Squadrons were a different breed. It was not unusual to hear about someone who was promoted in one of the squadrons to wake up, tied to a chair, and dangling from a rope about twenty or thirty feet off the hangar deck. Sometimes if they didn't wake up, their crews would hose them down.

I know traditions change over time, some are "enhanced" or done away with completely. But I like to think that I experienced some of the *honorable* traditions during my short time in the Navy. Honor is something that stays with you for the rest of your life.

(28) 75 Years Ago: The End of WWII in Europe

Dear Dad,

If you were still alive today, you'd be close to 99 years old. It's hard to believe that 75 years ago today, you would soon be part of something only a few people alive would have experienced—the end of World War II in Europe.

I can't imagine what you and the other men of the 80th Division and Allies were thinking 75 years ago today, on May 5, 1945. Did you know the Germans would surrender in two days? Were there any rumors or did G-2 (Division Intelligence) know anything?

I'm sure rumors were running rampant. After all, the 80th Division landed on Utah Beach in early August 1944, and had been beating the enemy back for months through France, Luxembourg, Belgium, Germany, and were now in Austria.

Your April 30, 1945 letter home to your mom and dad read:

Germany
30 April 1945

Dear Mom,

We're having wonderful weather now. However, it's still a little chilly early in the mornings and at night.

I imagine everyone at home is quite excited and happy about the war news they are hearing nowadays. I'll admit that the German soldiers we are fighting now do not have the discipline and morale that they used to (with the exception of SS troops, as they are fanatical), but just remember that although the news sounds good, the war with Germany is not over and will not be over until every German soldier is either killed or captured.

I wrote you before that I was now commanding officer of my company. Of course it's a lot more work, but then they're my boys and I'd gladly give my life for any one of them.

I have a good friend, Captain Tom Moss, from Orlando in my battalion. He is the Artillery Liaison Officer. Anyway, the other day, he took me up in one of his observation planes and we flew all over the city of Nuremberg. The city's about the size of Jacksonville and there are not a lot of buildings left standing. I stayed in that town several days so it was nice to see it from the air as well as from the ground. There are many cities and towns that are like that both here and in France & Luxembourg.

The other day when we went into a town I walked into a house & spotted a bathtub. Out of curiosity, I turned on the faucet to see if it would work. Much to my astonishment, hot water came pouring out!!! I thought at first that I was dreaming, so I let the water keep on running & immediately jumped out of my clothes & into the tub. And sure enough the water was hot, so I just lay back & let it soak in. But then that's war for you—sometimes you can live like a king off the fat of the land & other times it gets a little rough.

I'm feeling swell, am happy & love you dearly.

Devotedly, Andy

# # #

Dad, from the letters you wrote home to your mom & dad and your sisters, and your WWII diary, you described in great detail many of the battles you and your men bravely fought, from Utah Beach, through France and your Baptism of Fire at Argentan, through Luxembourg and the Battle of the Bulge, through Germany and what you found at the horrendous concentration camps, and finally in Austria.

Your letter, dated May 7, 1945, read…

Austria
7 May 1945

Dear Mom,

I know you're anxious because you haven't heard from me in about a week, but I'm getting along swell & have been doing a lot of traveling. I am now "somewhere in Austria." It is really a beautiful country with streams, rivers, lakes, hills & mountains. The mountains are a part of the Alps.

I image everyone at home is quite happy over the war news. And I hope & pray that an official declaration of peace will be made before you receive this letter. Thousands of Germans are giving themselves up. Yesterday, believe it or not, I took two machine-guns and a section of mortars & went into a town & persuaded 800 Germans & Hungarians to surrender themselves! I hope the rest of these devils will be as easy to persuade. One captain surrendered his entire company intact. After talking with him through an interpreter, I found that I had fought against his outfit back in Normandy. I was glad to see him without his tanks.

This is a hunter's paradise. Plenty of deer & quail. Surely wish dad could see this place.

I'm feeling swell & am happy.
Devotedly,
Andy

Dad, there is such a contrast between the letters you wrote home and the details in your diary. I know you didn't want your parents to worry—I've seen that same contrast while researching several wars.

You and the rest of the 80th Division returned home in January 1946 after remaining in Europe as an Occupational Force. Even though you knew you weren't coming home for a while (I'm not sure anyone back then knew when that would be), but you'd mentioned that since the end of the war, you were able to finally stay in billets with a roof over your head, your kitchen was with you for hot meals, and there was no more fighting and killing.

I know you lived with these war memories the rest of your life; I think all soldiers carry those with them. Sometimes they can talk about them, other times not. I know you chose not to talk much about the war when Anne & I were young.

You left us too soon at the early age of 68. But you left a legacy with us and I think writing your diary helped you bring closure to one of the most exciting, challenging, and tumultuous times in your life. You never knew this (well… *maybe* you do, now), but your WWII diary was published in 2005: ***You Can't Get Much Closer Than This: Combat with the 80th 'Blue Ridge' Division in World War II Europe.***

Say hi to Mom. We're all doing well: kids are fine, grandkids are great. We're fighting a different war down here right now—one we can't see. But we'll overcome and beat it. It's only a matter of time.

I'm feeling swell & am happy.
Devotedly,
Andy (your son)

P.S. Anne (my sister) told me that when you first got home from the war, your mom took you to buy new clothes because your old civilian clothes didn't fit. You felt like a new man. But your mom was also a good cook. I understand that your "new" clothes didn't fit after a couple of weeks. I can't compare my Navy service with your Army service, but I do remember having to buy new clothes when I came home from the Navy in 1977.

(29) May 8, 1945:
"Ours is a Quiet Celebration"

Today is Friday, May 8, 2020. Normally, I post a "Navy Friday Funny" to help blast you into the weekend. But today, I'm changing things up. There's a reason… stay with me here.

Today is "VE-Day," otherwise known as Victory in Europe. 75 years ago on May 7, 1945, Germany surrendered to the Allies, bringing an end to a devastating war. The next day, May 8, 1945, "Victory in Europe" was declared. There were celebrations all over Europe and the United States.

Celebrations *were* planned this year (2020) all over Europe to commemorate the end of World War II in Europe. But because of COVID-19, those celebrations have been postponed; understandably so.

I wanted to honor my dad and the courageous men who fought and also those who died during WWII. I can't help but think about what my dad wrote in a letter to his parents on this day, 75 years ago. "*Ours is a quiet celebration.*" The letter (and the text) is included below.

So perhaps you can take a few moments to *quietly* celebrate this day, this "Victory in Europe" day, to honor your fathers, grandfathers, brothers, uncles, cousins, and family members who fought in World War II.

# # #

From You Can't Get Much Closer Than This: Combat with the 80th 'Blue Ridge' Division in World War II Europe, by A.Z. Adkins, Jr., and Andrew Z. Adkins, III. Published by Casemate Publishers.

The following are my dad's words…

During the middle of the morning on May 8 we held up in a little country town in the vicinity of Spital, Austria, and got the men into billets. It was a beautiful day and the grass in my yard was green and soft. Slingshot (the company dog) and I were sitting under a tree and I was cleaning my rifle. Slingshot had oil all over him. I took my shoes off. It really felt good to wiggle my toes in the grass.

Sergeant Albert Melcolm came out and said there was a conference called for all COs. He read a message to us and said that an order would be down in a few minutes. I prayed to God. I asked 1st Sgt Ed Turner to assemble the company. When the men came in, I told them to sit down. Then I read General Eisenhower's order announcing the end of the war.

Supreme Headquarters Allied Expeditionary Forces
Eisenhower's General Order

1. A representative of the German High Command signed the unconditional surrender of all German land, sea, and air forces in Europe to the Allied Expeditionary Force, and simultaneously to the Soviet High Command at 0141 Hours, Central European Time, 7 May 1945, under which all forces will cease active operations at 0001 Hours, 9 May 1945.
2. Effective immediately all offensive operations by Allied Expeditionary Forces will cease and troops will remain in their present positions. Moves involved in occupational duties will continue. Due to the difficulties of communication there may be some delay in similar orders reaching the enemy troops, so full defensive precautions will be taken.
3. All informed, down to and including Divisions, Tactical Air Commands, and Group and Base Sections and Equivalent. No release will be made to the press pending an announcement by the heads of the three Governments.

Eisenhower

No one said a word for several minutes, then one of my men got up and said, "Lieutenant, read us that again, please."

# # #

Letter from my dad, Lt. A.Z. Adkins, Jr. to his parents in Starke, Florida.

Austria
8 May 1945

Dearest Mom & Dad,

Shortly after I wrote to you day before yesterday I received the cease fire order. At that particular time I had my shoes & shirt off & was playing with a little dog on the grass of some Austrian's yard. My men were all in houses taking it easy. My battalion had momentarily stopped in a little mountain village. I told the first sergeant to assemble the company.

As my men came marching up a big lump formed itself in my throat because many familiar faces were missing from the files of men who were to hear me read to them General Eisenhower's order that hostilities had ceased. I told my men to sit down & take it easy & that I had something to tell them.

Then I read to them General Eisenhower's order telling of the unconditional surrender. When I finished no one said a word. Finally, one man said, "Lieutenant, read that again please."

The day that we had died & bled for so long had finally arrived. No one knows what the word peace means except those who have been at war. As yet I feel no great emotional change. But gradually I am beginning to realize that there will be no more suffering & no more dying & the sensation is truly wonderful.

Tonight I am in another mountain village high in the Bavarian Alps. I have a radio & can listen to the celebrations that the people in England & America are having. Here, we are having a different type of celebration. Ours is a quiet celebration. We still have to maintain order, but we are so happy & it's hard for us to realize this mess is over.

I love you both dearly.

Devotedly,
Andy

(30) *USS Kitty Hawk* @ Bremerton Shipyards, Part 1

I proudly served in the US Navy for four years. Actually, one day shy of 4 years, only because I joined on a Monday and got out on a Friday, but that's another story.

Besides boot camp and ABH "A" school, I had two assigned duty stations: NAS Agana, Guam and *USS Kitty Hawk*. Yet, I feel like I had three "tours." Let me explain…

Back in my day, ABHs (Aviation Boatswains Mate-Handler) had a 4-2 rotation. That meant you'd spend four years on sea duty, then two years shore duty. Even though NAS Agana, Guam was land-based, the Navy considered it "sea duty." I don't know if the rotation is still the same or not. Do you?

Puget Sound Naval Shipyard

After Westpac '75, we were scheduled to spend a year in the shipyards. *Kitty Hawk* departed San Diego on March 8, 1976, and on Friday, March 12 we entered the dry dock at Puget Sound Naval

Shipyard in Bremerton, Washington.

This would be a one hundred million dollar complex overhaul, scheduled to last just over twelve months. There would be major work done on the ship, including renovations to accommodate the new, heavier F-14 "Tomcat" fighter jet and the S-3 "Viking" anti-submarine jet. These aircraft would replace the F-4 "Phantoms" and the S-2 "Stoofs" we had with us during Westpac '75.

We would also be in the dry dock and that was something I wanted to see—an 80,000-ton aircraft carrier on concrete blocks. The only vehicles I'd ever seen on blocks were some old Chevy and Ford pickup trucks back home in Gainesville, Florida.

While most people think that renovating a Navy ship, a REALLY big one, would be boring (they're right), I tried to make the most of it. There were several "firsts" for me: seeing snow for the first time; riding in a helicopter… over the snow-capped Olympic mountain range; and taking a college course that would shape my destiny.

How Can I Forget?

Chipping paint with needle guns… grinding and sanding bulkheads to bare metal… one or two coats of primer… four or more coats of gray paint… and the endless bitching and moaning of being stuck on a ship for a "nine-to-five" job. Believe it or not, that may have actually prepared me for some of my other future jobs. See "Hurry Up & Wait!" in an earlier blog.

Yeah, it wasn't as exciting as round-the-clock flight deck operations, but when you're in the military, you follow orders and do the best you can with what you have. THAT is a trait I think *all veterans* carry with them throughout their lives.

A New Adventure Awaits

I'd never seen snow in my life except in pictures and movies. Coming through Puget Sound on a bright sunny morning was spectacular. It was cool & crisp, the type of day my mom used to call a "Champagne Day." When we rounded the corner and spotted Mt. Rainier in the background—all 14,000 snow-covered feet—it was nothing short of spectacular. I knew I would enjoy my time here.

It took another few hours to maneuver *Kitty Hawk* through the sound and into the dry docks. The shipyard workers ("yard birds") would start pumping out water from the dry dock that day. It took two days for them to completely pump out all the water. *Kitty Hawk* was now resting on concrete blocks.

Our normal workday was 0730 to 1700 (5:00 p.m.). During the first two weeks in the shipyards, we were slowly getting used to the new work hours. Even though we may have finished our workday early, we couldn't leave the ship. The XO put the word out that "no one would leave the ship in civilian clothes before 1700."

Move to the APL

The shipyards provided several barges called APLs (Auxiliary Personnel Lighter) that had been outfitted to be two-story berthing compartments to house the ship's crews while their division berthing spaces were being renovated. The APLs were tied to the pier and located a couple of hundred yards away from *Kitty Hawk*. Another APL served as the mess hall.

You couldn't stay in your berthing compartment during it's renovation. It was quite a mess and would take time, so that's why everyone moved to the barge. Not every division aboard ship was being renovated at the same time, so berthing aboard the APL was on a rotation schedule.

Beginning the Overhaul

I know there was a lot more going on than what I personally saw (*Kitty Hawk* was a BIG ship), and others may chime in here (comments welcomed), but I can only share my perspective.

For V-1 Division, we would renovate all the V-1 Division spaces, including our berthing compartment, each of the three Fly compartments, and the Crash compartment. In addition, we would renovate the areas around those spaces, including the catwalks.

By renovate, I mean we stripped the paint down to bare metal and repainted with primer & paint (several coats). We also stripped the old vinyl tile off our berthing compartment floors and installed new tile (we did this more than once).

Those renovations would take a while, but there were about 80 of us in the division and we had an entire year.

About two months into the overhaul, our division officer purchased our own sand blaster to quicken the process. It took a lot of time and elbow grease to remove paint off the ship using needle guns (those *still* ring in my ears and I'd know that sound anywhere—anyone else?). Grinding and sanding down to the bare metal took about the same amount of time. Next, we sprayed on an initial coat of primer paint, then three coats of my favorite color—battleship gray.

At the *Bottom* of the Dry Dock

In May '76, the ship's Public Affairs Officer arranged for a tour to the bottom of the dry dock. I wasn't going to miss this. *Kitty Hawk* was sitting on hundreds of blocks each about four foot by four foot square and anywhere from three to six feet tall, depending on where they were placed under this ship.

The ship's two anchors were on the ground up toward the bow, along with hundreds of links of chain. Only one of *Kitty Hawk's* four propellers had been refinished. It was about twenty feet in diameter and had a nice shiny brass finish. We also walked under the ship's two rudders, each about fifteen feet long and about twenty-five feet tall. I felt my 6'4" frame was a tiny comparison.

My good friend, Glenn Law, reenlisted that day at the bottom of the dry docks. It was a proud occasion for Glenn—this was his *first* reenlistment. Neither of us realized it at the time, but Glenn would make the Navy a career for 22 years.

Finishing Up

Toward the end of September, the shipyard workers had finished scraping, priming, and painting the entire ship's hull. I can't tell you how much paint they used, but it must have been at least a million gallons. The next phase was to take the ship out of dry dock. They started flooding the dry dock, slowly at first to be sure nothing shifted too fast.

I was on board and it was quite an event to witness. The "great flood" took a full day. The tug boats then pulled us out of the dry dock into Puget Sound and parked us alongside pier #6. We would remain there until we left the shipyards.

In January, the yard birds began to apply non-skid onto the flight deck. We had already stripped it down to bare metal, applied a primer coat of paint, as well as some deck gray paint. Now the task was to spray on the non-skid. Two yard workers applied it simultaneously, so we furnished a few guys to help.

The only thing our guys had to do was to cover the flight deck "pad eyes" so they wouldn't get sprayed with the non-skid coating. They used a four-foot pole with a five-inch round metal plate attached to the end. I told them, "Don't wear your good shoes," since this stuff would splatter everywhere and on everything.

Watching the flight deck get back into shape was a sign we were getting close to finishing the overhaul in Bremerton.

It's *Never* Good Enough

Just to let you know how ridiculous the Navy can be "sometimes," we had to strip and paint our berthing compartment on *Kitty Hawk* four times and retile the floor twice. It seems there was "friction" between the two V-1 Division chiefs—one would give us a task to complete, then the other would complain and we would do it again. That's the primary reason I didn't make the Navy a career. I found myself caught between chiefs too many times.

Am I the only one?

I may have only been assigned *two* duty stations in my brief Navy career, but I consider the year in the Bremerton Shipyards another. It wasn't as exciting as flight ops or a Westpac cruise, but I learned a great deal.

Because we only worked during the day, I enrolled in Olympic Community College and took a course in electronics that shaped my next "journey." Long story short, I earned two degrees in Electronics Engineering (BS, ME). But that's another story for another day.

One more thing… after I graduated from the University of Florida in 1982, my wife and I moved to the Seattle area for a year. I worked for Weyerhaeuser Company (Federal Way, WA). We experienced many more *magnificent* views. On a clear, crisp "Champagne Day," this is truly "God's country."

And yes, we did visit Bremerton again and I ran into an old V-1 Division shipmate: Jack Kuiphoff. He had moved to Port Orchard, WA. We shared some great times and great memories.

(31) *USS Kitty Hawk*, CV-63 Decommission

After forty-eight years of faithful service, *USS Kitty Hawk* was decommissioned in Bremerton, Washington on January 31, 2009. Her active duty covered the Vietnam War, the Gulf War, and the Global War on Terrorism.

I was fortunate to have spent 2½ years aboard her ('75-77). Like many of you, those were times & memories I'll never forget. It's also one of the reasons I continue to write about *Kitty Hawk*—so much rich history before, during, and after my time. And like I wrote earlier, I love military traditions.

Throughout her lifetime, *Kitty Hawk* had 407,511 arrested carrier landings and 448,301 launches (who keeps these statistics?!). On August 6, 2008, the final trap and catapult launch was made, and the ship pulled into San Diego the next morning for the turnover to the *USS George Washington.*

She had been to the shipyards many times for retrofits, including the $100 million overhaul in Bremerton, Washington in 1976. She completed the overhaul in March 1977 and departed the shipyard on April 1 of that year to return to San Diego. I was aboard

Kitty Hawk during that entire time. I posted an earlier blog I wrote about the Bremerton Shipyards.

Last Ride from San Diego

Kitty Hawk left San Diego for her last "cruise" on Thursday, August 28, 2008 bound for its next stop: Bremerton Naval Ship Yards. The *Kitty Hawk* Veterans Association had worked diligently with *Kitty Hawk* and the Navy to get former shipmates on board for this last cruise and had been successful—a "Tiger Cruise" like no other.

I tried like the dickens to make that voyage, but I was stuck in Arkansas attending a trial in which I was to testify as an expert witness, but never did (that's a whole 'nother story, but "my side" won).

However, 65 former crew members made the cruise, including 38 who had been onboard when *Kitty Hawk* was originally commissioned in 1961; these sailors are affectionately known as "Plank Owners."

Decommission Ceremony

Over 2,000 former shipmates, crew members, guests, and distinguished visitors attended the January 31, 2009 decommissioning ceremony in Bremerton. I wasn't going to miss this opportunity. It was the first time I had seen *Kitty Hawk* in over thirty years.

Awesome Memories

When I first arrived in Bremerton, I walked up and down the pier several times, admiring *Kitty Hawk* and remembering my time aboard. Since the ship was docked on the starboard side as always, the bow was closest to the parking lot.

I thought of the times I poked my head out the mooring lines portholes, looking down at the huge keel piercing the water like a hot knife through butter. I looked up and caught sight of the safety nets and thought of the times I had stood on the forward flight deck at sea with my foul weather jacket unzipped, holding it with my outstretched hands and leaning over so far it felt as if I might fly. I wrote about those "bow flying" experiences in an earlier blog.

I thought of the hundreds of times I had climbed the steps and made the trip across the After brow saluting the colors and the watch, coming aboard and leaving the ship. The aircraft elevators were as big as I remembered—enough to hold two F-14 Tomcats with crews ready to man them for flight operations.

But because there were so many civilians and guests aboard, we couldn't tour any of the ship nor go up onto the flight deck. The ceremonies were held on the hangar deck. That was OK, though. Just seeing her and being able to board her was enough.

Emotions...

As I climbed aboard for the ceremonies, I experienced a wide range of emotions. In July 1977, all I wanted to do was get off the ship and get out of the Navy. Thirty-two years later, I was happy to be aboard her one last time.

The ship's CO, Captain Todd A. Zecchin, served as the master of ceremonies and gave an eloquent presentation of *Kitty Hawk's* rich history. Admiral Timothy Keating, Commander, US Pacific Command, also gave a wonderful, heartfelt speech—he had also served aboard *Kitty Hawk*.

I wish I could have visited some of my old onboard haunts, like the flight deck (remember all those hours of flight ops); Flight Deck Control (and the old "Ouija board"); the V-1 Division berthing compartment (and find my old rack); the mess decks (remember my favorite meal, *Kitty Hawk* "steak"); the Forecastle (stick my head out the porthole to look down at the massive keel); the Crash compartment (remember all those times gathered around the coffee pot); and the ship's onboard radio (entertainment) station, KRAL (and my "call sign:" Country Chet Adkins).

Imagine how disappointed I was when I found out later that several attendees asked on-duty sailors to "escort" them around the ship after the ceremonies. Had I known that, wow… a trip down memory lane. But alas, that wasn't meant to be.

Kitty Hawk's Future

Currently, *Kitty Hawk* sits in Bremerton, Washington, awaiting her last voyage to the scrapyard. There have been several efforts over the past few years to save her to serve as a museum, but those efforts seemed to have hit the wall.

The last I heard was that she had been sold to International ShipBreakers in Brownsville, Texas and awaits her tow date from Bremerton to Brownsville, which is yet to be determined.

Active Aircraft Carriers

The United States currently maintains eleven aircraft carriers in service. The newer "Nimitz-class" nuclear powered carriers cost about $8.5 billion (2012 dollars) to construct compared to the $265 million price tag for *Kitty Hawk* in 1961.

Of the eleven carriers, ten of them are Nimitz-class carriers, which means they are CVN-68 or higher. Since the Nimitz was first commissioned in 1975, the Nimitz-class carriers have been around for over 45 years. The newest class is the "Gerald R. Ford" class. The following is a list of active (as of 2020) United States aircraft carriers:

- (CVN-68) *USS Nimitz*, May 3, 1975—West Coast
- (CVN-69) *USS Dwight D. Eisenhower*, October 18, 1977—East Coast
- (CVN-70) *USS Carl Vinson*, March 13, 1982—West Coast
- (CVN-71) *USS Theodore Roosevelt*, October 25, 1986—West Coast
- (CVN-72) *USS Abraham Lincoln*, November 11, 1989—West Coast
- (CVN-73) *USS George Washington*, July 4, 1992—East Coast
- (CVN-74) *USS John C. Stennis*, December 9, 1995—East Coast
- (CVN-75) *USS Harry S. Truman*, July 25, 1998—East Coast
- (CVN-76) *USS Ronald Reagan*, July 12, 2003—West Coast (Yokosuka, Japan)
- (CVN-77) *USS George H.W. Bush*, January 10, 2009—East Coast
- (CVN-78) *USS Gerald R. Ford*, July 22, 2017—East Coast

Question?

I don't know if this matters now, but I'm sure someone has an answer.

I was aboard *Kitty Hawk* in the 70s and worked in Crash & Salvage, part of the V-1 Division. At that time, we only had 14 Crash crew members and had to maintain a *minimum* of 11 Crash crew on the flight deck during flight ops. That often meant a lot of long hours and at times, rotating double-shifts, especially during round-the-clock flying. More than once, several of us worked a 36-hour shift, but that's another story for another day.

When I spent a few awesome days aboard *USS Ronald Reagan* in 2009, as research for my book, there were 25+ members in Crash. But if memory serves, back in my day, there were 26 active aircraft carriers; now there are 11. My guess is that there are enough ABHs in Crash now with fewer carriers.

I don't know what the minimum requirement for Crash during flight ops is these days. Anyone know? If you know, please answer in the comments section; I'd greatly appreciate it.

That's today's weekly report (i.e., blog post). Stay safe, my friends.

(32) Memorial Day—2020

I didn't know him personally, but he was my father's best friend during World War II. They shared many hard times together, but also a few laughs along the way. Such is the way with combat veterans. My father never talked about Kad, but he wrote about him in his diary. After searching for many years, I found his family and let them know that Kad was a hero to my dad.

I didn't know them personally, but they were from my

hometown and attended Gainesville High School, graduating a few years before me. Several times a year, our local Vietnam Veterans of America chapter gathers at their graveside to honor these fallen friends: Marcus, Freddy, Alton, and Tommy.

I didn't know him personally, but I was there when he died on the flight deck of *Kitty Hawk* during Westpac '75. It's not something one forgets. During an aircraft crash, you have to stay focused and do your job to help keep everyone else safe, dealing with the emergency as best you can. I know many of my veteran friends know *exactly* what I'm talking about. You react, based on your instinct and training. Andrew was just doing his job, the same as the rest of us, but he was in the wrong place at the wrong time. He was a fellow shipmate.

I didn't know him personally, but I knew of him. He was about my son's age. I know his parents—we attend the same church. When I ride or walk by his memorial on the church grounds, I always pause for a moment and say a quick prayer. Chris was an Army infantryman who was killed in action in Baghdad. *Greater love has no one than this, that he lay down his life for his friends.* [John 15:13]

I don't know about you, but as I grow older, Memorial Day has greater meaning to me. There's a deeper feeling of patriotism, a deeper emotion of people who have gone on, and a deeper respect for those who served (even though they're still with us).

I can't help but think of my father, A. Z. Adkins, Jr., who served with the 80th Infantry Division during WWII, fighting in Europe. He made it back, but so many of his buddies didn't. I think of Dr. Rufus K. Broadaway who served with the 82nd Airborne and dropped in behind enemy lines the early morning of June 6, 1944, landing in an apple tree. He made it back, too. I married his daughter.

When I was younger, I know Memorial Day was important to my father, and I'm sure to Becky's dad, too. As a kid, while our parents wanted us to know about the war, they didn't want us to worry. I'm sure they spent a little quiet time on Memorial Day thinking about those men they served with who didn't make it back.

My mom once told me that before she met my father (they met after the war), she had written letters to a lot of soldiers who were friends from high school and college. Many didn't come back. I'm sure she also spent a little quiet time on Memorial Day.

This year, 2020, we are in a different, unprecedented situation. Normally, there would be thousands of ceremonies around the country, remembering loved ones, friends, colleagues, and unknown soldiers. However, because of COVID-19, those "normal" ceremonies won't be normal.

Instead, there will be different types of ceremonies, but we will remember—we *will always* remember. Veterans and families will still be out there, paying tribute to their fallen brothers, family members, and even strangers who gave their all.

I know I always take time during the day, like I've done for the past 40+ years, to remember the sacrifices our soldiers & sailors made in order for us to preserve our freedoms. I hope that you'll find a little quiet time to remember and *Never Forget.*

All gave some, some gave all.
Fair Winds and Following Seas.
We Have the Watch.

(33) Launch the Alert 5!

Not many people in this world have heard the phrase, "Launch the Alert 5!" I mean, why would they? Unless, of course, they watched *Top Gun*. But for those of us who have lived it… it's more than a catchy phrase!

After boot camp and ABH "A" School, my first duty station was NAS Agana, Guam. We worked two sections, "Port" and "Starboard," 24 hours on and 24 hours off. While on duty, five of us spent two, four-hour shifts a day in a cramped Crash firetruck on an "Alert Spot" between parallel runways. Our job was to wait for an aircraft emergency, or worse… a crash. While I was in Guam, I never heard that phrase.

On the Flight Deck…

It wasn't until I was aboard *Kitty Hawk,* that I heard it for the first time. I was in V-1 Division/Crash during the Westpac '75 cruise and worked on the flight deck. I learned what "Launch the Alert 5!" meant with the help of Crash Bos'n Ty Robuck, Crash LPO Jeff

Atteberry, and Asst Crash LPO "Big Willie" Williams.

Back in my time ('73-77), we were still in the "Cold War" era. Geez, that seems sooooo long ago, doesn't it? Among other things, the Russians constantly flew reconnaissance missions over our Carrier Task Force. We didn't trust them any more than they trusted us.

The two main Russian aircraft that buzzed us were *Russian Bears* and *Russian Badgers*. BTW, the Russians *still* fly reconnaissance missions over our naval forces.

Russians in the air

Once we left Hawaii and on the way to the Philippines, we established the "Alert 5." This meant we had two F-4 Phantoms on the catapults ready to launch from a cold start within five minutes. We kept this routine up around the clock during that two-week transit, primarily because the Russians didn't work nine-to-five either.

A pilot and RIO (Rear Intercept Officer) sat strapped in the F-4 cockpit waiting; nearby the starter equipment was hooked up to the aircraft, and a full plane crew, including maintenance and ordnance men. A Yellow Shirt director would be there, too, all ready to go at a moment's notice.

Launch the Alert 5!

If the Russians came *anywhere* near us, the Air Boss would announce over the 5MC [flight deck PA system], "Launch the Alert 5!" At that point, the plane captain and crew would start the F-4 Phantoms, spooling up the jet engines, checking the ordnance, and readying the aircraft for launch. The Yellow Shirt would direct the F-4 onto the catapult. Then, when all was ready, the Shooter would launch the aircraft.

The goal was to have at least one aircraft in the air within five minutes. Most of the time, we made it with time to spare, though sometimes there would be an equipment malfunction in which case we would launch the second one.

One pilot relayed to me that the usual scenario was that they would fly escort with the Russian *Bear* or *Badger* and smile and wave. Both the American pilots and the Russians would take photos of each other. But the pilot was always ready with his trigger finger, just in case. I was glad nothing ever happened; at least no one ever told us anything ever happened during those friendly exchanges.

And no, as far as I know, no pilot ever flew "inverted" over a Russian aircraft like "Maverick" did in *Top Gun*. Cheesy, but effective for Hollywood.

The "Alert 15"—Always a backup

We also had an "Alert 15" as backup. Two additional F-4 Phantoms or A-7 Corsairs would be ready to launch within 15 minutes after the Alert 5. The crews were *not* in the cockpit for the Alert 15, but close enough to get ready quickly in case one of the Alert 5s could not launch. It was spectacular to watch this remarkable teamwork in action.

I've been known to QUIETLY say, "Launch the Alert 5!" It usually happens when I'm driving and somebody recklessly pulls in front of me. Am I the only one?! Or, when I'm in the grocery store and the "new normal" one-way aisle is blocked. Yes, I often get a strange look or a harsh word. But it is *so* worth it.

Teamwork is the key

While I can easily recall the phrase, the thing I remember the most is how we worked together as a team to make it happen. Think about it. A couple dozen guys, many who don't know each other personally, but recognize the importance of what we do and why we do it. We constantly trained for this and when the time came, we executed.

After I left the Navy, I became a firefighter with the City of Gainesville—a "natural" progression for me as an ABH. Similar to those emergencies we faced on the flight deck, training kicked in when needed. There were times I worked with other firefighters, same city, but a different shift from a different station. But because of our continuous and intense training, we all knew what to do when we needed to do it.

Those instincts… still there

I'm currently retired, at least today. Who knows about tomorrow? But there are times, few and thankfully, far between, when I've come upon an emergency and my reactions mimic what we did in the Navy *and* the fire department. I react to the situation and hopefully, am of some help. Traffic accidents, house or car fire, or health emergency—I do what I can. I'm sure I'm not the only one here.

My wife is an RN (retired) and more than once we've come across an accident and worked the scene as a team. Like most first responders, that instinct to run *toward* the fire/accident/emergency is still there.

I do miss the challenges and the excitement of the flight deck. Yes, it was dangerous, and we lost a few crew members. We can Never Forget them. But we were there for a mission. We were there for the Navy. We were there for our country. And that, my friends, makes me a proud Navy veteran!

(34) No Better Place to Die

La Fiére Causeway, Normandy—76 Years Ago

I can't begin to describe what I've been through the past few days. Hell, what we've all been through. I'm not even sure where to begin, it's all been such a blur. Nonstop… day after day, hour after hour, minute by minute. It hasn't let up.

We've lost hundreds, if not thousands, of men—good-hearted young men, fighting for freedom, helping to rid the world of this tyranny. I knew many of them; trained with them all; and led a few through these past horrific days. It's hard to believe they died. I hope not in vain. Time will tell.

This place we're at today—and have been for the last couple of days… it's just a bridge across a small river (Merderet River, I believe). I'm not even sure the name of this bridge, if there is one. We're a few miles west of Sainte Mère-

Église. They've had some heavy fighting, too. But our mission today is to capture and hold this bridge. I know we'll succeed—we're Airborne.

We're next to a French manor. There's a group of buildings that overlook the east end of the bridge. There are a half-dozen farm buildings, two or three stories high, and constructed with heavy stone walls. There's also a five-foot wall surrounding the buildings. It's probably strong enough to withstand mortar fire.

Today is June 9, 1944. *Just a few days ago, we parachuted into Normandy, behind enemy lines. It was pitch black, about one in the morning.*

We were supposed to jump the day before and were ready and waiting inside our C-47 for what seemed like hours. Me and 17 other men in my stick. But at the last minute, they called it off, "Not today."

Judas Priest. What a bloody letdown. We'd trained for two years, built up our bodies and our minds, and were strapped in with 100 pounds of gear. 6,000 of us in 830 C-47s, ready for our greatest challenge. Now, we had to wait… again. No sleep for many. But we're the 82nd Airborne; we're paratroopers; we do what we're told.

We jumped the next day in the early morning hours of June 6th—D-Day. I could've died then.

I vividly remember being called up to the front (I was in the rear of the C-47). The pilot yelled out, "Lieutenant, the plane has been hit, and it's not responding. I've lost formation and I don't know where I am. We're going to be over the Channel in a few minutes. If you're going to get out, you'd better get out now!"

I shouted to the Jumpmaster on my way back to the rear of the plane, "Let's go!"

While a normal jump would have been about 800 feet off the ground, for some reason, this one was only about 300—my chute opened and immediately, I landed in an apple tree.

Fortunately, I was only a few feet off the ground. I cut myself out of the chute and dropped down—nothing broken. I still had my carbine and my pistol. I was ready…

Just when I thought I was safe, I heard a faint, "click." I searched my pockets, but couldn't find my "cricket" signal clicker. I heard it again. I know he was one of us, we'd trained for this. But I couldn't see him. I took a cautious step, crunching a few leaves. "Halt, or I'll shoot!"

"Don't shoot. I'm Lieutenant Rufus Broadaway of the 507th. I lost my clicker."

Who's our CO?"

"Captain Bob Rae."

"Damn, Lieutenant! I almost shot you." It was my platoon sergeant.

That was close—one of many close calls to come. Thank the good Lord, I'm still here. And to think, if I'd listened to my mother, Miss Eva, I'd be a Baptist preacher.

We were scattered all about, not sure where we were, or who else was around. Did anyone else from my stick land nearby? It was still dark, but luck was with us. There was a little moonlight.

Sergeant Hobeck, our platoon sergeant (the one who challenged me earlier), and I made our way toward our mustering area and we ran into other paratroopers. Some were from our company, others from other units.

I didn't realize it at the time, but the area we were supposed to land (our drop zone) had been flooded by the Germans. Many of our troopers landed in water over their heads. With all the extra weight, they didn't have a chance and drowned.

The next day, we ran into Captain Rae, my commanding officer. Instead of heading to Sainte Mère-Église—our original objective—he told us we were needed to help men from the 325th Glider Infantry Regiment at the La Fiére Causeway. It was an important bridge to capture so our landing troops coming from Utah Beach could cross the Merderet River and advance inland.

By the time we got here, the 325th soldiers were fighting hard against units of the German Seventh Army, who had taken up defensive positions at the end of the bridge. Our soldiers were bogged down and not able to get across.

They were planning to mount another assault. Our small group of paratroopers approached from the southeast. Word came down from Brigadier General James Gavin (Assistant Commander, 82nd Airborne Division) to take the bridge, "at all costs."

The bridge is only a two-lane road, maybe 20-30 feet wide, probably 3/4 mile long. The Germans had flooded the Merderet, so the river was actually wider than usual.

There's a hole in the manor fence wall alongside. That's where we'll head, then out onto the bridge and attack. But there's a German heavy machine gunner who has zeroed in on that opening. I can tell because there are a lot of dead and dying soldiers nearby.

But we've got to go—attack and hold the bridge "at all costs." We're Airborne. We will succeed. My heart is pounding and sweat drips from my brow.

The air is full of lead, almost indescribable. The stench of cordite hangs in the air like a heavy fog, burning my nostrils. Mortar fire, heavy machine guns, light machine guns, rifle fire—it's all around and the smoke and haze have made it nearly impossible to see what lies in front of us.

German soldiers are in foxholes on both sides of the river, throwing their potato masher grenades. Dead men are scattered about—both theirs and ours. It's a pure bloodbath and there's only one way out.

I'm in the middle of this. No time to think. If I get hit, then I get hit. I can't worry about that now. I've—we've—got a mission. The success of the D-Day invasion is counting on us.

The Germans are still firing at us, but we're making headway. We're slowly pushing them back and gaining ground.

There's a narrow farm road on the left, sunken with hedgerows. I run and jump over a fence at the end of the bridge. Next to me is another young lieutenant; he joined us in England, before the jump. I don't even know his name.

We both stand on our toes, trying to peek over the embankment to see where the mortar fire is coming from. He falls back into my arms; a shot right through the single lieutenant bar on his helmet.

I'm 23 years old… the war just got serious for me.

That was 76 years ago, today (June 9, 2020). Dr. Rufus K. Broadaway was one of the greatest men I've ever known. He was a man of integrity and a man who wasn't afraid of anything. He'd faced death many times during the war and afterwards on the operating table—he was a top-notch and compassionate surgeon.

After the war, he returned home to his bride, Marion, and their daughter, Judy, whom Rufus had yet to meet. She was born July 2, 1944.

After finishing college (he started at Mississippi College *before* the war and ended with a degree from Tufts University, then a medical degree from Harvard Medical School), he, Marion, Judy, Dana, and little Becky moved to Miami, Florida. Rufus was one of the first faculty members at the *new* University of Miami School of Medicine.

I first met Rufus in 1980—I drove down from Gainesville to ask his and Marion's permission to marry his youngest, Becky. He wasn't a tall man (I'm 6'4"), but I looked up to him from the start. I knew he'd fought in WWII, like my own father. Both men, to me, were bigger than life.

Becky and I married in 1981. Over the years, he and I had several opportunities to talk about the war. It obviously affected him, as it did all combat soldiers. I had the privilege to conduct an Oral History of Rufus in 2010. Amazing man, amazing stories.

Also, here's a link to a short 6 1/2 minute video of Rufus describing the D-Day jump and the battle for La Fiére Causeway. This was recorded in 2009 by his good friend, Dick Martin at Oak Hammock, Gainesville, Florida.

We lost Rufus a few years ago at the age of 95. Marion, 99, is still with us, still vibrant as ever, and hunkering down in quarantine, like the rest of us.

During these current trying and troubling times, we can Never Forget the sacrifices these courageous men made in order to preserve the freedoms we continue to enjoy today. God Bless all our veterans.

Notes:

1. "No better place to die"
    a. Attributed to Captain John J. "Red Dog" Dolan, a paratrooper CO. He wrote to his men, "We hold here. There is no better place to die."
    b. The title of a book, written by Robert M. Murphy, 82nd Airborne.
    c. The name of an upcoming movie (pre-production), directed by Captain Dale Dye.
2. The battle to secure the La Fiére Causeway has been described by historians as "the bloodiest small unit struggle in the experience of American arms."
3. More than 250 American troops died in the battle for the La

Fiére Causeway.

4. More than 830 Douglas C-47 aircraft took part in the Normandy D-Day drop.
5. The Manoir de la Fiére, located next to the La Fiére bridge still stands. Becky & I have been there and met the owners. Rufus & Marion had been there a few years before. The family welcomed us all.

(35) US Navy "A" School

For those of you *not* familiar with the military, once you've completed boot camp, your next training "stop" is a trade school, specific to your "job." I wrote an earlier blog about boot camp.

"A" school is a term used in the Navy. Marines call it "School of Infantry." Army goes to Advanced Individual Training. I don't have a clue what Air Force does (yes, that's a dig at one of my *two* Air Force friends).

Lakehurst, New Jersey

After boot camp, I went to ABH A School in Lakehurst, NJ. Lakehurst is probably best known as the site of the famous blimp Hindenburg disaster when on May 6, 1937 the German zeppelin *Hindenburg* caught fire at Naval Air Station Lakehurst. The ABH A School moved to NAS Pensacola, Florida in October 1996. That's where all ABHs now go for training.

I arrived in Lakehurst on September 6, 1973. At the time, the ABH rating only had an A school, which lasted about six weeks.

Other Navy ratings have "B" and "C" schools, which were more technical and required longer training. Since this was in the 70s, ABHs were rates, or "billets," for men. That changed while I was still in the Navy, though we did not have women on ships at that time.

The standard Navy issue definition (1973) of an Aviation Boatswain's Mate-Handler (ABH) was to "operate and service aircraft ground-handling equipment and machinery, operate and service aircraft crash, firefighting, and rescue equipment, handle aircraft afloat and ashore, and perform crash rescue, crash removal, and damage control duties."

Basically, we moved aircraft around on the flight deck of an aircraft carrier; I was also a Crash firefighter, on Guam and aboard *Kitty Hawk*. Both were "perfect" choices for me.

What did I learn in "A" School?

Well, number one, I learned to "drink like a sailor." Stay with me here…

I joined the Navy when I was 18 years old. I had a year of college at Florida State University before enlisting.

I didn't drink when I was growing up in Florida. In boot camp, there was no time. So when I got to Lakehurst—the drinking age was 18 in New Jersey—I quickly discovered an Enlisted Men's Club close to the barracks.

More than once, I "stacked my drinks," which was legal back then. For the *new* generations, that means that you can order drinks *before* finishing them. It wasn't unusual to find 4, 5, or 7 drinks in front of me. Best of all, they were only a dollar apiece, which fell within my "liquor intake budget." My E-2 salary (1973) was $342.30 a month.

More than once, I was grateful that the barracks were located *downhill* from the EM Club. That way, someone could roll me down the hill after I had one too many. Which, BTW, happened "one too many" times. But that's another story. I'm sure there were "races" downhill at one time or another. But I can't remember.

Temporary Holding Company

When I first arrived, there weren't enough men for a full training class. Similar to arriving at boot camp, I was assigned to a temporary outfit. There were about a dozen barracks at NAS Lakehurst, but only about half were renovated enough to live in. During the first couple of weeks, command assigned those of us in temporary companies to help renovate the barracks.

My job was to hang ceilings. There were four of us working diligently as a team on this particular complicated task. The ceilings were *supposed* to be level all around the room. But let's just say that some were a little more level than others, depending upon how long I had stayed at the Enlisted Men's Club the night before.

ABH "A" School Class Starts

After two weeks, we *finally* got enough guys in to start the class. ABH A School lasted about six weeks. Most of our class work consisted of book learning, which was pretty boring to me, and practical exercises for ABHs. I did fairly well in most of the class work, scoring high on most of the tests. But I *really* liked the practical exercises, especially firefighting. I don't know why, but it seemed natural to me.

The course instructors started us out slowly on the firefighting skills for obvious reasons. It was awesome, though, when you got to fight a big one.

Initially, they would pour about fifty to a hundred gallons of gasoline on an old jet in the fire pit and light it. Our job was to put out the fire. Now and then, they would hide a 150 pound dummy, affectionately named "Oscar." We'd have to find him and pull him out of the fire without damaging him. You can bet he'd seen better days.

Walking Through Fire

At the time, Crash firefighters wore asbestos proximity suits, silver coated with a big hood you often see in movies. I loved walking directly into a wall of flame in that suit—it made me feel invincible.

Did it get hot while in the fire? Not really; there was an insulated liner inside the suit which helped keep the heat out. The outside coating reflected the heat. The helmet also had an asbestos hood with a face plate coated with gold flake. The gold was there to reflect the heat, so while you could walk directly through flames, the face plate would help reflect the heat from your face.

After a few minutes though, the suit would begin to heat up. But the idea is that while you *could* walk through direct flames, you shouldn't stand around in them toasting your tidbits. Our job was simple—put out the fire and rescue people.

Teamwork is always the Key

It all boiled down to the teamwork of the crew. At the school, performance was graded as a team. There were five of us in the crash truck: the driver, the turret man, the hose man, and two rescue men. One of the instructors drove the fire truck—none of us had been trained. I'd learn to drive them in Guam.

Each of us has a job to do. The driver got us close to the fire. The turret man knocked down the initial flames with a flame-retardant. The hose man blazed a trail (pun intended) through the remaining flames so the two rescue men could drag Oscar out. The hose man then put out the rest of the fire. Pretty simple, but it took teamwork.

The Infamous "Dream Sheet"

We all got our orders for our next duty station in mid-October. When I first got to A School, we filled out paperwork called a "dream sheet." The idea is that *you* would tell the Navy where you wanted to go for your next assignment and then they would send you wherever *they* wanted you to go. I filled out the West Coast for *ship* duty and the Philippines for *shore* duty.

Not that I wanted to get that far away from home, but because I had heard the west coast had the better cruises. My orders read, "Report to Naval Air Station Agana, Guam." I was into scuba diving so my new duty station would be a diver's dream. Now if I could just figure out where in the hell Guam was.

I graduated from A School on Halloween, October 31, 1973, and had a few weeks leave before I left. I didn't know how long I would be stationed on Guam, so those few weeks were extra special, spent with family and friends.

What Did I *Really* Learn?

I learned how to fight fires, something I would do again in civilian life. I learned how to drink and more importantly, how *not* to drink. But to be honest, I came away with a greater sense of self-respect, which certainly boosted my confidence—another step along in my "journey." I also learned the importance of teamwork. You cannot do it alone, you have to rely on others to help.

Not too many people in this world dare to get close to a big fire, let alone walk through it. Most of the time when there's a fire, people run *away* from it, as well they should. Firefighters and first responders run *toward* it. And that instinct still sticks with me today.

(36) Gone in an Instant

Trigger warning: *This blog post describes the death of a shipmate on the flight deck of an aircraft carrier. It doesn't contain specific details, but it may trigger flashbacks for some—I hope not. My intent is to share with you an experience that I hope no one will ever have to face. But for those of us who served in the military, it happens.*

This has been a difficult post to write. To tell you the truth, I started to write it several times over the past few months, but backed off for various reasons. One of which is our current COVID-19 crisis. My intent of "A Veteran's Journey" blog is to always keep a positive outlook on life.

But I needed to finish this one. Something recently happened that brought this subject closer to home. Life is precious. You just never know…

Most of us have experienced death at one time or another during our lives. It may have been a loved one or it may have been a close friend. For us veterans, we may have experienced death during our tour of duty.

I don't think we get to choose when or where—that's left to a higher authority. But after losing someone close, we must find a way to keep moving forward.

The Most Dangerous Job?

It's been said that "working on the flight deck of an aircraft carrier is the most dangerous job in the world." That *may* be true, but I think it depends on your personal perspective. I'd venture to guess that soldiers facing an enemy on the battlefield, shooting at you with an intent to kill, or hunkering down in a foxhole, trying to avoid an artillery barrage, may beg to differ.

There's an old adage, "Whether active duty, discharged, retired, or reserve, a Veteran is someone who at one point, wrote a blank check made payable to The United States of America for an amount of up to, and including, their life." To me, THAT.IS.HONOR.

So in my humble opinion, it doesn't matter if you're Army, Navy, Air Force, Marines, or Coast Guard. When you sign on that dotted line, you join an elite brotherhood, knowing that your life *may* be on the line.

Westpac '75

My second duty station was *USS Kitty Hawk, CV-63* (my first was NAS Agana, Guam). I was a third-class petty officer assigned to the V-1 Division and worked on the flight deck in Crash & Salvage. During Westpac '75, we lost three sailors. I don't know the specifics of two, but let me tell you about the one that happened on the flight deck.

Our overseas home port was PI (Philippine Islands). We were in and out of PI seven times in 1975. The Vietnam War has "officially" ended a few months earlier, but we were still on alert during our deployment. On this particular trip, *Kitty Hawk* headed out to sea for a two week exercise—standard *at sea* time for us.

The first aircraft off the deck for flight ops was the "plane guard" helicopter from the HS-8 ("Eightballers") squadron. We had a 2-man Crash tractor manned on the flight deck, which was required for pre-launch. The rest of the Crash crew were in the Crash compartment on the starboard side of the island or standing outside shooting the breeze, waiting for flight ops to begin.

Crash to the Flight Deck!

Suddenly, the Air Boss announced over the 5MC (flight deck PA system), "Crash to the flight deck! Crash to the flight deck!" We all hauled butt out onto the deck; all we could see was a helicopter trying to take off. It took a couple of seconds to realize what happened.

The starboard landing gear had collapsed and one of the helo ground crew was dangling from the wheel well. The main wheel tie-down chains had been removed, but the helo was still attached to the deck by the tail wheel tie-down. The pilot was hovering over the deck at about six inches. The Plane Captain was directing the pilot to keep steady, but it was tough since the helo was still chained to the deck.

My pal, Buddy Laney, was suited up on the back of the Crash tractor and witnessed it all. He told me later that it was one of the most harrowing experiences in his life, being so close to a killer helo and not being able to do anything about it. He said his training kicked in and they were preparing for the worst—a crash on the flight deck.

Several of us grabbed a rescue stretcher and dunked down, inching closer to the helo. About that time, the helo dropped the ground crew member. We grabbed his legs and pulled him out of range of those dangerous rotors. Yes, we were under the spinning rotors.

The flight deck medic was right there with us trying to save him, but this shipmate was long gone. We put him into the stretcher and carried him over to the #2 elevator where he was lowered down to the hangar deck and rushed to Sickbay.

Meanwhile, the helo was still hovering just a few inches off the ground. I heard that our Crash LPO Jeff Atteberry was on his stomach crawling toward the helo.

The only way this thing could end safely was to release the chain.

When Jeff reached the tie-down, the Plane Captain saw what he was doing and helped the pilot keep the helo steady. Once released, the pilot flew off and everyone on the flight deck breathed a sigh of relief.

Recovery & Salvage

We weren't done yet—the second part of our name was "Salvage." We had to figure out how to get the helo and its crew safely back on board.

Fortunately, part of our Crash arsenal included a "crash dolly," a heavy metal dolly about three feet by four feet with four heavy rubber wheels. Several mattresses were brought up from below decks and tied to the crash dolly. We then anchored it to the flight deck in the recovery area using a half dozen tie-down chains.

In order to get the helo safely back onto the flight deck, a squadron Plane Captain served as the director—standard operating procedure. We parked a tow tractor in front of him so that if something happened, he could at least duck down behind the tractor. We moved all nearby aircraft to the aft part of the flight deck—no need destroying or damaging aircraft if it could be helped.

We also had the Crash truck parked on the starboard side with a full crew. Steve Deaver, a big strong Texan who had recently joined Crash in the Philippines, was fully suited up and standing on the back of the crash truck, ready to roll. I know his heart was pumping.

All unnecessary personnel were cleared off the flight deck. V-1 Division and squadron personnel were in the catwalks standing by with fire hoses in the event of an all-out crash and subsequent fire(s).

As the helo approached, the Plane Captain directed him slowly over the flight deck. The pilot hovered a few feet over the deck to let the crew and passengers jump down.

The pilot knew what needed to be done, but it took him three times to ease down before he could place the crippled starboard wheel safely onto the crash dolly. I can't say enough about the teamwork between the Plane Captain and the pilot. Both knew what *could* happen.

We had placed enough mattress pads on the dolly so that when the helo landed, it would be level. Once the full weight of the helo was down, he kept the rotors going until the Blue Shirts and helo crew could tie the helo down to the flight deck. When that was done, the pilot cut power to the engines. He was safe, we were safe, and there is a God.

R.I.P. We Have the Watch

That was one of the scariest and most challenging times I faced in the Navy. But we cannot forget that we lost a shipmate. He was pronounced DOA when he got to Sickbay.

From what we heard, he was removing the landing gear safety pins, something *always* done before a launch. The port side pin came out fine, but the starboard side would not budge. He started banging on the pin. It came out all right, but when it did, the wheel collapsed. He was inside the wheel well, so he was caught by the weight of the helo.

At the time all this happened, I didn't feel any emotions, just adrenaline, and my training kicking in. You bet, my heart was racing; I think everyone felt the same. I'll be honest though, when it was all over, I had to find a quiet place for a bit to decompress. I'm sure I wasn't the only one that day.

I've often thought about that airman over the years. The flight deck is a dangerous and sometimes unforgiving place to work. He was just doing his job, like me and all the others on the flight deck. It could've been any of us.

I thank the good Lord for that Plane Captain and the helo pilot that kept the aircraft in the air and steady, so we could do our jobs. We all worked together as a team with men we didn't know, but trusted. That's what you do in the military.

I didn't know AMS3 Andrew J. Wantulok, but his death on September 23, 1975 affected me personally.

This event took place almost 45 years ago and the description may be a little fuzzy. But I'm fairly certain it is accurate.

Gone in an Instant

A month ago, my wife's cousin was killed while riding his bike. He had been an avid cyclist for most of his life and traveled the world over with his wife and family. He easily had 100,000 miles of cycling experience under his belt. He was on an early evening ride, close to his home in Chico, California, when he was hit by an SUV. We don't know the details, but he was gone in an instant. He and his wife had just visited us in Florida a year earlier on their way to their new home and their new life. Dave had recently retired at age 65.

You never know when your time is up. Death leaves an impression that you carry with you throughout your life. How can it not? I've often said that "death is part of life's journey," and I truly believe that. But still… sometimes it hits close to home.

Sometimes I think about the "what if…" I mean, am I the only that does that? *What if* the helo had hit the deck, crashed, and sent shrapnel flying all around. There's not a lot of room on the flight deck and plenty of personnel around. Then… I remind myself that it did not happen. That keeps me from wandering down a potentially dangerous path.

I've been married for almost 40 years to the love of my life, with two wonderful kids, and two energetic grandkids. I worry about them more than myself. I think most parents and grandparents do. Don't get me wrong, I want to live long and prosper.

My first two "long life" goals are to live longer than my dad (68) and my mom (72). I think I'll make it. Both of them died of smoking-related cancers. I don't smoke.

My next "long life" goal is to live long enough to see my daughter's *third* grandchild graduate from college. While she's married to a wonderful man, they don't have any kids yet. I've got my fingers crossed. And before you ask, yes, she's heard me say that before.

Lofty goals? Absolutely, but certainly doable. Right now, I'm like most of you—taking necessary precautions to avoid the coronavirus. That's my day-to-day goal. So far, so good.

(37) Hawaii, One Word: Beautiful!

Do you remember your first time? You know, your first liberty in port? I bet you do. You *never forget* your first time.

My first duty station was NAS Agana, Guam, a "land-based" station. I blogged about it earlier.

My *second* duty station in my short, 4-year Navy career was *USS Kitty Hawk, CV-63. Kitty Hawk* was an aircraft carrier (decommissioned in 2009 after 48 years of service) and I was honored to be a part of her crew from March 1975—July 1977.

When I arrived aboard *Kitty Hawk*, I was a young, 20-year-old ABH3 (Aviation Boatswains Mate-Handler, Third-class Petty Officer), assigned to V-1 Division, Crash & Rescue. We worked on the flight deck.

Westpac '75 began in the latter part of May 1975. *Kitty Hawk* left San Diego and was in and out of Hawaii twice prior to departing for the Philippines, our overseas home port.

This was the first time I'd spent more than just a few hours in Hawaii. While I had flown through Honolulu on my way to Guam,

I only experienced the inside of the airport.

The first time in port, we had two full days of liberty. What does a sailor do while in paradise? Hint: surfing & drinking would be a couple of options. What a surprise!

I wanted to learn how to surf, so several of us from V-1 headed to Waikiki Beach. The water was warm and crystal clear, and the scenery was absolutely stunning. I thought Guam was gorgeous. But to me, this was *much* better. In one word: Beautiful!

Surf City!

I was out in the Hawaiian surf with a rented surfboard, not having a clue what to do. But then I was Navy—this wasn't the first time, nor the last, when I found myself in a predicament of "not knowing what I was doing." But I was a water skier, so this *should* be easy, right?

My buddy pointed out a very nice-looking young lady who was surfing like a pro. He told me he'd talked with her a little earlier and she seemed quite nice and may help me learn to surf. "Her name is Karen."

I swam over to her and with my standard Navy issue confident and cocky attitude, I said something like, "Hi. Your name's Karen, isn't it?"

She said, "Nope, it's Jennifer." Damn, I fell for that... again. I apologized and then left with my tail tucked between my legs, or whatever it is when you're on a surfboard.

After a little while longer, I got the hang of it [pun intended] and learned to surf with the help of a new friend. I really enjoyed surfing—it "freed my soul," so to speak.

The day was hot during the peak surf time, and the sun was beating down on my back. I hadn't put on any sunscreen, but I didn't care. I was born and raised in Florida and spent many summers in the sun, both at Kingsley Lake and at the beach. Yeah, I'd pay for that one later.

What Bar Are We In?

After a glorious day of sun and fun at Waikiki, our next stop... bar hopping. I don't recall the name of the bar, but I remember that

I was drinking something blue and there was a girl dancing on our table.

Drink prices were sky high—$3.50 for a cocktail. Back in Gainesville (Florida) the same drink would have only been $1.50. After a couple of those drinks, the night got a little blurry.

We stopped by another bar, the Red Door Lounge, which had a band playing country music. This was more like it (not that I didn't like women dancing on my table, mind you). Several of the guys egged me on stage to sing with the band, so... why not?

The band welcomed an addition—another sloshed sailor. I managed to get through *The Auctioneer* easily, then the lead guitar player handed me his guitar and I stumbled through *Wildwood Flower*. I tried to sing *Rednecks, White Socks, and Blue Ribbon Beer* as a tribute to my fellow shipmates, but I was pretty far gone, so I quit in the middle of that one.

Everybody in the bar was laughing, probably because I was making a fool out of myself. It was *not* the first time and wouldn't be the last.

USS Arizona Memorial

The next day, we visited the *USS Arizona* Memorial. It was officially dedicated on Memorial Day 1962. We took a ferry to get out to her. The memorial actually straddles the sunken battleship. It was quiet and serene, like a library, or a church before a funeral.

Standing on the platform overlooking her, she was not as big as I originally expected, at least not when compared to *Kitty Hawk*. The most visible feature of the ship is the barbette to gun turret number three—it pokes out above the water.

I remember looking down into the crystal clear water and thinking of all those men, more than 1,100 still entombed. It was a very solemn time for all of us. They were Navy... we were Navy.

As years have passed by, I've reflected on that visit many times. These courageous sailors were *our* age and lost their lives thirty-four years earlier defending our country, their ship, and their fellow shipmates in an unprovoked attack.

As we often say in the Navy, "Rest in Peace, shipmate. We have the watch."

I would visit the *USS Arizona Memorial* again years later and experience the exact same feelings. Some things never change.

Eject, Eject!

After our *first* port call in Hawaii, we sailed out on May 29 for a few days of training exercises. During flight ops on the first day, an F-4 Phantom from the VF-114 squadron (Aardvarks) was conducting touch and go landings on the carrier.

I was in my usual place on the flight deck foul line next to the crash tractor. We always kept our eyes on the aircraft coming *and* going, just in case we noticed something wrong and could anticipate our next action.

During one of the touch and go's, the jet's horizontal stabilizer locked and the Phantom went vertical. We watched the pilot apply full afterburners. But with a nose up attitude and minimal speed, the aircraft stalled.

The pilot and RIO (Rear Intercept Officer) punched out at about eight hundred feet. *Davy Jones* claimed another Navy aircraft with no chance of recovery. Fortunately, the two crewmen parachuted into the ocean to our starboard side. The SH-3 "plane guard" helo (HS-8 squadron, "Eightballers") was there within a few minutes, recovering both crewmen with no injuries.

I don't know what happened, but we later learned that it was an "equipment malfunction." That's the Navy's way of saying, "it's none of your effing business." That was OK with me—at least they didn't crash on the flight deck.

General Quarters!

We left Hawaii the second time on Wednesday, June 4. Our next stop would be the Philippines, a trip that would take about two weeks.

A little over a week out, *Kitty Hawk* experienced a *major* problem. One of the main engine rooms flooded around midnight. A steam line ruptured followed by a sea valve malfunction, causing the machine room space to flood. I believe everybody down there escaped with no injuries. That was the *good* thing.

The *not-so-good* thing? The CO announced over the 1MC, "Anyone with scuba equipment report to engine room #1… *and* bring your equipment." That was scary.

We went to General Quarters. V-1 Division personnel moved all crash equipment and aircraft to the port side on the flight deck since the flooding caused the ship to list to the starboard side. *That* was a "fun" night.

But no one died.

I was reminded that during an earlier cruise, Westpac '73, *Kitty Hawk* lost six sailors to a fire in the No. 1 main machinery room. Another 38 were injured fighting the fire. I wasn't there, but I read besides losing men, the ship remained at General Quarters for a day and a half.

The flight deck is *not* the only dangerous place to work on a ship.

I've mentioned this before: we don't know when our time is up. Make the most of what you can when you can.

I say that "responsibly," but then, I'm working on another blog titled, "Here, Hold My Beer!"

(38) Hook, Line, & Sinker

Happy Anniversary to me!

Today, Friday, July 10, 2020, is an important date for me. Forty-seven (47) years ago, I signed on the dotted line and enlisted in the United States Navy on July 10, 1973 for a 4-year commitment. But because I joined on a Tuesday and was discharged on a Friday, it was *only* 3 years, 11 months & 29 days… but who's counting?

I began a 4-year *journey* that would impact the rest of my life. Like many of my Navy veteran friends, I bought into the complete Hook, Line & Sinker story and *never* looked back. I only questioned that decision a few times (i.e., during boot camp). I'm sure I'm not the only veteran who feels that way. I blogged about my Orlando RTC boot camp experience earlier.

My "Navy resume" is short. But instead of a Ph.D., I *earned* a DD-214. I am a proud US Navy veteran.

- ABH2 (Aviation Boatswains Mate-Handler, 2d Class Petty Officer)
- *USS Kitty Hawk*, CV-63 / V-1 Division: March 1975—July 1977
- NAS Agana, Guam / Crash & Rescue: November 1973—February 1975
- ABA "A" School, Lakehurst, New Jersey
- Orlando RTC (Recruit Training Center)

It all started when I Flunked Out of College

"Dismissed for Academic Reasons."

There it was on paper in black and white; May 1973. Florida State University asked me to depart their wonderful institution. I started college when I was 17 years old, but that was my first year away from home and my grades were not up to par. I couldn't believe they kicked me out because of my 0.7 G.P.A.

I knew my Mom and Dad would be pissed. I'd been up in Tallahassee for almost a year. They paid the entire way, both for school and a "little more." I *thought* I was doing all right.

Of course, it took me a few weeks to get used to being away from home. And sure, since the drinking age was eighteen, I partied a little—OK, I partied a LOT. But that wasn't an excuse for F.S.U. to flunk me, was it?

Scuba Diving?

I was into scuba diving and worked at a local scuba dive shop, North Central Florida Diving Center, owned by my friend and scuba instructor, Barry Kerley. We dove the wrecks and jetties off Panama City, but my favorite spot to dive was the St. Marks River, right off the bridge in Woodenville.

I knew there was work to be done at school. The only problem? It obviously wasn't that important to me.

Army… or Navy?

Back then, a thousand thoughts ran through my mind. I could always join the Army and follow in my dad's footsteps. He served with the 80th Infantry Division and fought the Germans in WWII. That's probably what he might want me to do, even though we were still fighting a war in Vietnam.

I talked with my friend & fellow diver, Ed Kelly, about his great times and adventures in the Navy. Since I loved to scuba dive, there would be more opportunities to dive with the Navy than with the Army.

I borrowed my roommate's bike the next day and rode down to the local recruiting office in downtown Tallahassee. I can't remember the recruiter's name, but he was very cordial, informative, and seemed trustworthy. He, of course, saw me as one more for his monthly quota though I didn't feel that at all during the visit.

Aviation… what?

After we talked a bit, he gave me a thick book of Navy jobs to look through. So much to read with so little time. Since I didn't have a long attention span, it didn't take me long to point to the job I wanted.

Similar to spinning a globe and stopping it with your finger, I said, "Here's what I want to do."

Actually, my thought process was very logical. I loved scuba diving, so that brought me to the Navy. I loved being outdoors, and I had always been interested in airplanes. As a kid, I probably assembled about every jet and airplane model on the shelves. They fascinated me.

I selected the rate, Aviation Boatswain's Mate-Handler. "No problem," the recruiter said. "The Navy needs ABHs and you will probably be able to go anywhere in the world you want." I'm sure he said that to everyone that walked through his door.

Since I was eighteen, I could legally sign documents. "Sign on the dotted line. Press hard, the third copy's yours." And with that simple gesture, I was now a member of the United States Navy.

Many people have asked over the years if I signed up for the

Navy to *really* join ("it's not just a job, it's an adventure") or if was I worried about being drafted into the Army. Even though the Vietnam War was winding down in 1973, we still registered for the draft.

Honestly, I *wanted* to join the Navy. The thought of being drafted never crossed my mind.

Mom, Dad… I have some "news"

A few days later while working at the dive shop, my mom and dad called. My dad said, "Son, we just got your report card and wanted to talk with you about your grades."

"Yeah, I know. I didn't do as well as I should have."

Before my dad or mom could say another thing I blurted out, "I joined the Navy." There was a silence—a *long deafening silence.* I heard a collective gasp from the other end of the phone. My mom said, "What?" I told them again.

I forgot whatever else they said during that phone call, but my mom told me years later that she and my dad were proud of me at that moment. They said that I had made one of the most important decisions in my life. More importantly, I stuck with it, all Three Years, Eleven Months, and 29 days.

Question for you… What drove *your* decision to join the Navy? Add to comments section below. We all have a story to tell.

(39) Good Mornin' to ya, *Kitty Hawk!*

Back in the mid-70s ('75-77 to be more precise), *Kitty Hawk* had several entertainment "options" while at sea. KTTY was the onboard TV station. Television shows included live daily news updates, movies, and periodic "events," broadcast on the closed-circuit televisions, located in all the main compartments.

Real-time Flight Ops on TV?

Another *Kitty Hawk* TV channel displayed real-time flight deck operations. If memory serves, there were two "views:" one view captured aircraft launching off the bow cats (Cat1 and Cat2) and the waist cats (Cat3 and Cat4).

The other camera was the PLAT (Pilot Landing Aid Television), a camera located in the flight deck, smack dab in the middle of the recovery area. The PLAT providing a view to watch recovering aircraft.

Most of the training films we watched were from the PLAT and flight deck cameras. Many were horrific crashes that I really don't care to watch again, but at the time, they did help us learn what to watch for (like an aircraft coming in too low or a pitching deck; or an aircraft catching a wire, but applying full throttle and full elevator only to slam down onto the deck).

Not great for Saturday night movies, but when I look back on it, these training films taught us to constantly stay alert. I've said this before and I'll say it again: the flight deck is one of the most dangerous places to work. Anything can happen at any time. You have to always keep your head on a swivel, a habit that's stayed with me my entire life.

I don't know if anyone ever bet on this, but I'd "heard" scuttlebutt that *some* crew members would place bets to see if a landing aircraft caught a wire (i.e., recovered) or "bolted," meaning they missed catching any of the arresting gear cables and took off again. I cannot confirm or deny any of this.

Kitty Hawk Radio

Kitty Hawk also had three onboard radio stations: KROC (rock), KSOL (soul), and KRAL (country). These were 24-hour stations, and the DJs were sailors from all different walks of life, each with a 2-hour time slot.

BTW, a little side note, just in case you were wondering… In the United States, radio & TV stations *east* of the Mississippi River start with a "W" designation, like WKRP & WDVH. Stations *west* of the Mississippi start with a "K." I believe this started back in the 1920s when radio & TV first began broadcasting. There are exceptions (aren't there always?), but that's TMI for this blog post.

My DJ "Journey"

But besides working the flight deck, I had my own radio spot: 6a—8a, the Country Chet Adkins show.

"Chet" was my Navy nickname, "assigned" to me by John Melcher, a fellow Crash & Rescue crewman I served with while stationed at NAS Agana, Guam. I played guitar and sang country music and since my last name was Adkins… well, John thought that was appropriate. No arguments, here. I couldn't play guitar like my namesake, Chet Atkins, but at least I could spell my name right <grin>.

How did I become a DJ, you ask? I know you want to know, so stay with me here…

"Big Willie" was an ABH2 and the Assistant Crash LPO. He always had a cool head, even during flight deck emergencies. His actual name is Shelby Williams. He's still around… I think in Texas. Everybody called him Big Willie because… well, he was a big guy. He always watched out for us peons, but he also had an agenda.

He'd been around a while and knew that the FNGs were always short of cash right before we hit port. So if anybody needed any money (before payday), Big Willie was the man to see. Borrow $20 and pay back $40 on payday. No problem, cash was always available.

The Country Chet Adkins Show

Big Willie was also a DJ on the KSOL (soul) station and had a two-hour spot in the late afternoon. I thought that was cool, so I asked if I could do country. BW took me down to the station, introduced me to the station manager and a few days later, they put me on the air. My radio personality name (i.e., my "call sign") was Country Chet Adkins.

It took me a little while to get the hang of it, but I was a natural. The station equipment included two turntables to play LPs (*Long Play* records) and two eight-track tape players to play commercials, not that we had that many. There were hundreds of LPs. We even had a phone for shipmates to call in requests.

There wasn't a lot of room inside the three studios, but they were all located next to each other. But man… we all played a lot of great music.

I initially started out with an afternoon spot 1400-1600 (2:00 to 4:00), because I worked the night shift. Later, I changed to 0600-0800, which worked out well since we didn't need to be on the flight deck until 1000 for flight ops.

I always started my show with a Chet Atkins guitar solo: *Yakety*

Axe, an old Boots Randolph song (Boots Randolph played the saxophone and it was originally called *Yakety Sax*).

I would crank *Yakety Axe* up at 0600 and after a few seconds came over the radio: "Gooooood morning to ya, *Kitty Hawk*! It's six o'clock and time again for the Country Chet Adkins show."

This was years before Robin Williams' *Good Morning Vietnam* movie. When I had a rough day on the flight deck, it was very relaxing to get down to the radio station and go on the air early the next morning.

It helped me get my head right for the upcoming day. We only operated the stations while at sea. I kept my radio spot the entire Westpac '75 cruise and again after the Bremerton Shipyards.

Back in the mid-70s, country music was in a transition from *old* country to *new* country.

That sounds like a cliche (it is), but we saw the transition from the old steel guitars, mandolins, and twangy sounds to the more "modern" (that is, 70s modern) sounds that featured artists like Johnny Rodriguez, Tanya Tucker, Johnny Paycheck, Dolly Parton, Charlie Daniels, Lynn Anderson, and John Denver… and, many more.

Yes, country music is always changing. But then again, almost all music genres change over time. Face it, "the only constant in this world (especially now) is change."

DJs need to speak clearly, not 'southern'

I learned several things while DJing. First and foremost, I learned to speak clearly. See… I'm from the south and many of us have a special way of talking.

For one thing, we often make two-to-three syllables out of any one-syllable word. And we also appreciate our southern drawl. So I had to learn to enunciate my "spiel" (that *ain't* no southern word).

Back then, all of our music came on AFRN (Armed Forces Radio Network) record albums. And we'd have to queue them up, so we could talk during the song intro, but (like professional DJs) end our conversation at the exact time the singer would begin.

It's not as difficult as it sounds, because you not only had a stop watch in front of you, you also had the exact amount of "intro" time written on the record.

Those were some good times and happy memories. Maybe, just maybe, I and the other DJs aboard *Kitty Hawk* were able to provide some relief throughout the workday through the music we played.

Somewhere in my house—probably in a moving box—I have several cassette tapes that I recorded during my show. I'll find them sometime.... then, I gotta find a tape player to play them.

Music is still in my soul today. I continue to play guitar and sing country music, but I also picked up the mandolin and started playing with a country gospel band: WSU, "Whoever Shows Up." We've been playing off and on together coming on 30 years. We don't play that often, especially now during this COVID pandemic, but now and then, I'll pick up the guitar/mandolin and try to belt one out.

(40) Midrats… and other Navy Chow

How does one "accurately" describe Navy Chow?

Before I get into *that* discussion, I have to remind myself that I had another cook before the Navy—Mom.

See, I'm from Florida and my mom was a great cook. Lots of great vittles from her third-generation cast iron skillet. But biscuits… well, mom's few feeble attempts at biscuits turned into gut-bombs. Everything else was fantabulous, but I learned to make biscuits later in my short journey as a city firefighter—it was required!.

My dad, on the other hand… Well, he was good at cooking steaks and "square" hamburgers. Now and then, he'd empty the fridge and make what he called, "Garbage Soup."

Man, that was tasty, but my mom could never understand how soup stains appeared on the ceiling when he cooked. Anyway, that's the kind of cooking I was used to *before* the Navy.

An Introduction to Navy Chow

I had two duty stations: NAS Agana, Guam and *USS Kitty Hawk*. And yes, IMO, Navy meals served in Guam were different than those served aboard *Kitty Hawk*.

In July 1973 at Orlando RTC (boot camp), my first "introduction" to Navy Chow was SOS. I'll let you Google the meaning. But "Shit on a Shingle" was standard Navy-issue chow which, as many of my fellow shipmates will attest, "ain't that bad." Basically, SOS is creamed chipped beef on toast.

Have I indulged SOS since July 10, 1977 (my discharge date)? In a simple word, "no." And… I'll leave it at that.

Well… what about "Sliders?" Surely we can talk about them. Or "One-eyed Jacks" or "Barney Clarks." Absolutely. Sliders are hamburgers and One-eyed Jacks/Barney Clarks are sliders with a fried egg. And yes, over the years, I've been known to partake in these delectables from time-to-time.

No, I won't mention Baluts, because they weren't served in the mess hall (at least, in *my* time). For those who served on the west coast, I'm sure you're familiar with this Filipino "delight." For those who served on the east coast… you're lucky. And… I'll leave it at that.

I've said it before and I'll say it again. Navy chow was always plentiful and I *never* went hungry. I know some shipmates might argue, but while I enjoyed some meals more than others, I appreciated what we had when we had it… even those times when we couldn't get down to the mess hall.

Though there was that time during a work party when one of the crew "found" a package of frozen hot dogs, "Use before Dec 1956." And… I'll leave it at that.

Mess halls @ NAS Agana, Guam

We didn't have a galley at the Crash Barn and we couldn't drive the crash trucks over to the mess hall because it was too far away from the airfield. We got our meals on wheels, so to speak.

We had two sections in Crash, Port & Starboard. When the shift changed over, there was a signup sheet listing what was available for dinner that night.

About 1800 (6:00 p.m.), one of the guys would take a pickup truck over and load our chow, provided in stainless steel serving pans by the cooks, and drive back. We used standard Navy issue metal trays (same as aboard *Kitty Hawk*) and loaded up.

Mess halls aboard *Kitty Hawk*

On *Kitty Hawk*, there were three galleys, located on the second deck, two decks below the hangar deck. The main galley was located aft and served the main meals buffet style. Unlike today's Navy chow lines (serve yourself), back then the cooks and stewards would serve you when you put your tray out. We bussed our own trays so we just took them over to the scullery, dumped excess food into a can, and handed our trays, glasses, and utensils to the dishwasher.

A second galley, called the "Express Line," was located forward. This was similar to a "fast food" drive-in, serving mostly hamburgers, hot dogs, french fries, spaghetti, and whatever else the cooks could scrape up quickly. I ate a lot of meals there, not because I was a burger or dog fanatic, but because it was fast—with 80 guys in our division to feed and long times in the chow line, we needed to get something quick, choke it down, and get back on the flight deck to let another group come down.

Powdered Milk

We usually had whole milk when we first left port. However, after a few days, the galley ran out of fresh milk and began serving powdered milk. That tasted like [insert your own definition here], so it didn't take me long to learn to eat cereal *only* the first few days at sea. After that, breakfast for me consisted of *anything* other than cereal.

Chow lines

Chow lines at sea were always long. It was not unusual to wait for an hour to get chow. On an aircraft carrier, the galley needed to feed more than 5,500 men every meal (3,000 ship's company and 2,500 air wing). That's about 17,000 meals served a day, including more than 900 loaves of bread when the ship was underway. That's a lot of meals to prepare, cook, and serve.

Lines would wrap around and through several different passageways. I learned early on to bring a book while waiting for chow. Yeah, I read a lot of books in the Navy.

We ate in shifts, so to speak. During flight ops, we had to maintain a minimum crew of Yellow Shirts and Crash crew on the flight deck so each unit would send five to six guys down at a time. Most of us would get our chow, eat like the dickens, and then head back up to relieve another crew man.

And yes, I still eat "like the dickens," even though I've tried to slow down over the years. Old habits, I guess. Am I the only one here that still does that?

Saltines… a Sailor's Best Friend

Saltines were always readily available, and I ate a lot of them. I didn't know it at the time because I just liked Saltines, but for some reason, they were magic. During my time on *Kitty Hawk*, I never got seasick, which is more than I can say for a lot of other guys, especially the FNGs.

There were times when I would have a bag of greasy, buttered popcorn and see some new guy a little green and offer him a handful. He headed for the side of the ship, tossing his cookies. I know it wasn't a nice thing to do. And… I'll leave it at that.

Wardrooms

Officers had it made—they ate in wardrooms which, as I understand it, had round tables, tablecloths, real dishes, and better yet, a full wait staff. I never ate a meal in the wardroom. In fact, I never stepped in one until I spent a couple of days aboard *USS Ronald Reagan*, CV-76, while researching for my book. I wrote about that experience earlier.

CPO Mess

Another place I never had the "privilege" to visit was the CPO (Chief Petty Officer) Mess. From what I've heard, the CPO Mess had *the best chow* on the ship, even better than the Wardroom. The CPO Mess is also commonly referred to as the "Goat Locker."

I did not know this (I had to Google it), but the term "Goat Locker" dates back to the days of wooden sailing ships. Goats were kept aboard ships because they'd eat anything and always provide milk. Why we didn't have goats aboard *Kitty Hawk* after the first few days at sea (see above), I'll never know.

When the rank of Chief Petty Officer was established (before I was born), there were no separate berthing areas for these senior enlisted men. Rumor has it that they shared the same quarters as the ship's goats.

I'm sure if I'm not 100% accurate, a CPO will correct me, ask me to remove my cover (hat) and knock three times to enter. And… I'll leave it at that.

Midrats

Midnight rations (Midrats) were always available for those working late into the night. They were served in the forward galley from 2300 to 0100 (11:00 p.m. to 1:00 a.m.) and usually were pretty good. Many times, midrats were left-overs from the evening meal, or burgers, hot dogs, or eggs to order, and other "quick to fix" items. I ate there many times when I was assigned the night shift.

Does anyone else still use the term "Midrats"… for anything consumed after dinner, especially late-night snacks?

Gedunk

Also just forward of the main galley was the Gedunk shop. This small onboard store was like a small neighborhood convenience store or a New York Bodega and had an assortment of snacks and cigarettes. While I didn't smoke, some of my friends did and their standard line was, "Where else can you buy a carton of cigarettes for ten bucks?"

There were times when the chow line was too long for me to wait so I would head to the Gedunk and pick up a bag of chips or crackers, some canned cheese, and a can or two of sardines. It wasn't that healthy, but it *was* something to eat. Unfortunately, while standard Navy issue meals came with the job, the Gedunk shop did not and this came out of our pockets.

Steel Beach Picnic

I made one cruise aboard *Kitty Hawk*, Westpac '75. Great times and I've written about it several times on this blog. One thing we didn't experience during that cruise (at least I don't remember) was having a steel beach picnic.

From what I gather, they were held on the flight deck or in the hangar bay and usually after a long at-sea time period. The goal was to help improve crew morale.

For those who haven't heard the term, think of a "beach picnic," with grills for ribs, bratwurst, hot dogs, hamburgers, and whatever else the galley needs to cook before the expiration date. Usually, there was music (most Navy ships have several bands playing different styles of music), games (e.g., Frisbee, football, volleyball) and plenty of time to just "chill."

And yes, at times (from what I've heard), crewmembers were allowed to drink beer.

Soup from the Galley—Always Happy to Help

Several times during the Westpac cruise in the colder climates (Sea of Japan), I called down to the galley and asked if we could get some hot soup. They said sure, come on down. So a couple of us made several trips down several ladders to bring up a huge 25-gallon container of hot soup. It tasted so good and everyone in our crew got a break to come in and down some. It helped us choke down the cold sandwiches, too.

I wrote about our Westpac '75 cruise through the Sea of Japan earlier.

Coffee, Black, no sugar please

Both in Guam and at sea, we always kept the coffee brewing. That is something we never ran out of. That tradition continued when I joined the Gainesville Fire Department after leaving the Navy. There were many times when I would come in off the flight deck to warm up and to get a cup of hot coffee.

I started drinking coffee in Guam (I was 18 years old). There was always plenty of cream and sugar. But on *Kitty Hawk*, I usually was the last one in to get coffee (I always let my guys come in first), there was *never* any cream or sugar left when I got mine, so I had to drink it black. I got used to that and still today, I drink it black.

There were also many times when I would tip the 100-cup coffee maker to get the last dregs of coffee, complete with grounds. But in cold, miserable weather, even that coffee tasted good.

Roach Coach

No blog post about Navy chow can be complete without a mention of the infamous, "Roach Coach." All Navy sailors on both coasts are familiar with the Roach Coach.

For those not familiar with the term, a "roach coach" is similar to what's now known as the "food truck." Of course, back then, there really wasn't a "food truck health inspector." Back in the day, you could get a lot of different types of sandwiches and my favorite… Bean Burritos.

And… I'll leave it at that.

(41) Haze Gray and Underway!

I left the Navy in July 1977, more than 47 years ago. I wrote about that "anniversary" a few weeks ago in an earlier blog. I still remember walking across elevator #3 toward the after brow, carrying my seabag over my shoulder and my 11-string Greco guitar (that's *another* story for later) in my hand. One final salute (in civilian clothes) and off I went.

Over the years, my wife and kids had suggested several times that we take a cruise. I seemed to have always found a lame excuse: "I've been on one."

Finally, in 2008, my wife & I joined our son, his future wife, and her parents to take my first "non-Navy" cruise. It was the first time I'd stepped onboard a ship larger than a small fishing boat. And that first night I told Becky, "Why did I wait so long?"

That trip brought back many (MANY) memories for me. But I digress… this blog post is NOT about *that* cruise.

Haze Gray and Underway

This is a phrase you won't hear in the Army, Marine Corps, Air

Force, or Coast Guard… unless, of course, the Marines use the Navy as their personal *Uber.*

You won't hear submariners use this term either, though they may try to argue. It's a phrase typically associated with Navy surface ships, which BTW, are a "haze gray" color as opposed to the dark gray or black color of submarines.

I had been in the Navy for more than 1½ years; I was a 3d Class Petty Officer (ABH3); and I was well-versed in aircraft firefighting and had driven several different types of Crash firefighting trucks, all while stationed at NAS Agana, Guam.

But when I stepped aboard *Kitty Hawk* in March 1975, I was a FNG assigned to the V-1 Division / Crash. I worked on the flight deck, so I had to learn quickly what to do, when to do it, and where I should be, and more importantly, where I should *NOT* be (don't cross that "foul line"). Otherwise, I might find myself in a world of hurt.

We had a great Crash crew, led by CWO2 Bos'n Ty Robuck, Crash LPO Jeff Atteberry, Asst Crash LPO Shelby "Big Willie" Williams, and a dozen other sailors dedicated to their job. They all helped break me in to my new assignment.

Synchronized & Choreographed

V-1 Division had about 80 sailors; the Air Department, about 500. But the entire ship had 5,500 (3,000 ship's company and 2,500 air wing). I knew this at the time, but because I was caught up in my own "flight deck" world, I rarely had the opportunity to visit other departments on the ship.

It didn't dawn on me just how well synchronized a ship had to be to get underway—everyone doing their jobs in order to safely make it out to sea. But we all made it work.

I'm sure I'm not the only one, but if I could go back and make an announcement to the entire crew over the 1MC, I'd say this:

"Thank you all for what you do. You helped make my one and only cruise, Westpac '75, challenging and memorable—something I will Never Forget. I couldn't tell you back then, but I'm telling you now."

Saying Goodbye to Loved Ones

While I wasn't married, nor did I have a girlfriend, a lot of my fellow shipmates were, many with kids. Most had mixed feelings about heading out to sea—they hated leaving their families, but they loved being back on the ocean.

Not many people outside the military understand that complex lifestyle or those feelings. I think that's one reason I took section duty when we returned for a fellow V-1 Crash or Yellow Shirt who had family waiting. It just seemed like the right thing to do.

Heading Out to Sea

Not all shipmates would agree, but when you're in the Navy and accustomed to working aboard a ship (for me, an aircraft carrier flight deck), I loved heading out to sea.

The routines of getting underway, the tugboats, the BMs working the shore lines, the ship's bells and whistles, the Captain's announcements over the 1MC… it was almost like "leaving home to go home."

During the day, you could watch as the ship moved through the channel, past the other Navy ships docked in the bay, past Point Loma and the Old Point Loma lighthouse, and out into the deep blue sea.

The flight deck is about 60 feet above the water line, so we had great places to observe our surroundings from not only the flight deck but also the catwalks. Heading out, I'd usually be on the starboard side, watching as we passed the lighthouse. Heading back in to San Diego, I'd want to see North Island, so again, I'd be on the starboard side.

I know there were guys on the hangar deck looking out through the massive hangar bay elevator doors, guys on the aft sponsons catching a glimpse, and of course, all those perched in their spots on the island superstructure.

San Diego is the 2d largest surface ship base in the US (Norfolk is the largest). Even though *Kitty Hawk's* home port was North Island, we could still see the activity at the main base from the flight deck.

We Were *Never* Alone

There were a lot of ships home ported there and when we got underway, others in the fleet were already out waiting for us or about to get underway themselves following us. That was also a comfort, knowing there were other Navy ships in the carrier battle group accompanying us on the cruise. We were not alone—we were *never* alone.

There's something about standing on the flight deck, high above the water, feeling the cool breeze coming over the bow, breathing in the fresh ocean air, and just taking it all in. "The wind in our faces, the salt in the air, and the glory of another find day in the Navy." I miss those days.

Of course, there are other "things" I don't miss about the Navy, but let's keep this post positive, shall we?

Squadrons Flying Aboard

When we first left port, the flight deck was bare, except for Crash equipment, aircraft tow tractors and tow bars, and a few other odds & ends. No aircraft were on board yet, but they'd join us in the next couple of hours.

The CVW-11 Air Wing squadron maintenance & administration personnel were already on board. I can't remember how long it would be after we passed Point Loma, but the pilots would fly the aircraft from their various squadron bases (most in California) out to *Kitty Hawk*. I imagine it's the same routine in today's Navy. That would be the first recovery of the cruise and helped us "get our act together," so to speak.

Nighttime

If you haven't been on a ship, it's hard to describe the nights when at sea. But if you can imagine being on the beach late into the night with no campfires, no ambient light, 5,000 of your closest friends near you, and no alcohol… well, you get the idea. Beauty without booze.

We flew at night. The enemy never rests, why should we? Sometimes, we'd fly 24 hours around the clock, but not that often during Westpac '75.

At night, everyone needed to be more cautious, especially during flight ops. There are no flood lights on the flight deck. Aircraft follow Yellow Shirt director signals who use cone-shaped flash lights. You had to constantly watch yourself around the aircraft, even those parked and tied down. I've got more scars than I can count on my shins from tripping over those blasted tie-down chains in the dark.

When the Air Boss announced over the 5MC (the flight deck PA system), "Standby to recover aircraft," that meant it was time to go to work. During recovery operations, instead of landing aircraft every 20-30 seconds as we did during the day, they'd spread it out to 60 seconds on a clear night. Longer if it was cloudy or stormy. Those were some "interesting" times, too.

At sea, I tried to find time to embrace the serenity of the ocean after a full day and night of flight ops. Many times, I'd take in those brief glimpses up at the clear sky of millions of twinkling "horizon-to-horizon" stars in between recovering aircraft.

Then again, there was the opposite spectrum of calm, mirror-like water versus the raging storm-tossed water crashing over the bow, 60 feet above the waterline. In those rough seas, I'd look over at an accompanying tin can destroyer a hundred yards off the starboard side and watch its bow dip *completely* underwater while *Kitty Hawk* cruised over the tops of the waves and be thankful I was on *my* ship and not *that* one.

Cruisin' *After* the Navy

On my first cruise *after* I left the Navy—some 30+ years later in 2008—I knew exactly where I wanted to be when we left port. The bow!

No problem… there was plenty of room and only a few others up there. Feeling that slight salt breeze, smelling the clean air… the movement of the ocean… yeah, it was one of those happy "triggers" that immediately took me back to getting underway aboard *Kitty Hawk*. Well… except for a few things like no aircraft flight operations, no duty stations, no hollering CPOs, but plenty of booze. Yeah, I could get used to that.

What little hair I had was all wind-blown, but I didn't care. The salt spray was in my face and on my glasses, but I didn't care. I had to squint at the bright sun, but I didn't care. I was at sea, "Hair's Gray" and Underway.

Nights on the cruise? Well, let's just say that party goers are sometimes louder than an A-6 Intruder ready for launch on Cat 2. But I will say this—there was *more* alcohol on the Holland America cruise than on *Kitty Hawk*. And no tie-down chains.

I hope that in today's COVID world, you might be able to find that wind blowing in your face, see those millions of twinkling stars when you look up at the night sky, smell that fresh sea air and remember… those good times you had when you can proudly say at one time, "I was haze gray and underway."

(42) The Shipyards, Part 2

A few months after our Westpac '75 cruise, *Kitty Hawk* set sail for the Bremerton Shipyards, for a year-long scheduled maintenance period. This dry dock time would include a $100 million complex overhaul, beginning in March 1976 and ending in April 1977.

Bremerton, Washington is located on Puget Sound in the magnificent and breathtaking Pacific Northwest.

Prior to leaving San Diego, several shipmates crossed over to another carrier for the next Westpac. But after careful consideration, I chose to head north with *Kitty Hawk*. It was the right decision for me and while I didn't know it at the time, that path helped set my life's journey. I blogged about the onboard shipyard experience a few weeks ago.

Off Duty in Bremerton

Our normal workday in the shipyards was 0730 to 1700 (5:00 p.m.). V-1 Division had 6 sections, so we had section duty every six days. That meant we stayed aboard the ship that entire day and night. Many of us in the division were single, so staying aboard wasn't that big of a deal.

Bremerton is not well-known for its downtown nightlife. Of course, there were a few bars right outside the main gate, but that's probably true of most military bases.

Not that all men & women in the military drink—many of us also liked to play shuffleboard. You know those waist high tables, 2 feet wide, 20 feet long with plenty of saw dust. Yeah, *those* "indoor" shuffleboards. And darts… who didn't play darts in a bar?

New "Duty Station" Surroundings

It wasn't long before I settled into my "normal routine" for a *new* duty station: get acclimated to my new surroundings, find the best bars, drink cheap beer, and find good, cheap places to eat. Then, after a few months, wind down and stay onboard more, but take time explore the area. This would become my shipyard off duty routine.

That was my routine when I first arrived at NAS Agana, Guam and again, in San Diego while in port. Hong Kong and Japan were short visits, and I really did explore more of the area outside the bars. I just wished I'd done more of that in the Philippines, our overseas home port for Westpac '75.

And let me tell you, Puget Sound has many scenic drives and beautiful destinations. My good friend and fellow V-1 Division Yellow Shirt, Jack Kuiphoff, settled down in Port Orchard, WA after he left the Navy.

The weather in Bremerton, as throughout most of the Puget Sound area, is typically drizzly, dreary, and wet. But there are times when it is absolutely beautiful. On those cloudless, crisp, blue, "Champagne days," there is no prettier place on God's green earth.

Bill's Inn

While there were several bars near the main gate, I found "my place." An out of the way bar with a live band called, "Bill's Inn." I and several other shipmates frequented Bill's many times. Now and then, we'd stay until closing time. Why does Mickey Gilley's song, *Don't the Girls all Get Prettier at Closing Time*, ring a bell?

It was after one of those late night/early mornings that I learned a valuable lesson. When you stop at an all-night diner for midrats, it helps if they have pictures of the food. Yeah, you know exactly what I mean.

Tell me I'm not the only one who discovered that important tidbit of information. Go ahead… I'll wait.

Sunday nights at Bill's Inn was jam night. That meant anyone could get on stage with the band and jam with or without an instrument. I, "Chet" Adkins, was a regular there on Sunday nights. One night while jamming, I was in for the surprise of my life.

I was twelve years old when I got my first guitar. My sister, Anne, had given me a record called, "Play Guitar with the Ventures." The Ventures was an instrumental group with a ton of hits in the 60s such as *Wipeout*, *Pipeline*, *Hawaii Five-O*, and *Walk, Don't Run*. I literally learned how to play guitar with The Ventures.

I was on stage one Sunday night, jamming with the house band, when a tall, skinny guy joined us and took over the lead guitar. I was playing rhythm and belting out Hank Williams and Patsy Cline tunes while Nokie Edwards was playing lead behind me.

Nokie was the lead guitarist for The Ventures. I was beside myself. Here I was singing some of my favorite country songs and the lead guitarist backing me up was the guy who literally taught me how to play, though he obviously didn't know. It was almost like a dream.

We found a little time to chat, but being from the Tacoma, WA area, his fans swamped him. That was a night I would never forget the rest of my life.

Mount Rainier

I'm from North Central Florida—it doesn't snow here… well, except that one time when I was away… in the Navy.

Bremerton is only two hours from Mt. Rainier, a 14,000+ foot high gorgeous snow-capped beauty in the Cascades. I drove up to Mt. Rainier with a friend on January 1, 1977. That was the first time in my life I'd ever seen snow. I was *the kid* in the candy store, throwing snow balls, making snow angels, slipping and sliding every which way, and busting my butt several times. I had the time of my life. And to think it took twenty-two years for that to happen.

I would visit Mt. Rainier again—many times. After college,

Becky & I moved to Kent, WA to work for Weyerhaeuser Company, headquartered in Federal Way, WA. We spent a wonderful year in the Pacific Northwest. And yes, we visited Bremerton a few times, though it just didn't seem the same as it did in the mid-70s.

My first Chevy

I didn't own a car before joining the Navy. When I was in high school, my dad bought a 1954 Willy's Jeep: three-on-the-floor, NO power steering, and a hardtop. We used it mostly for hunting, but when I was old enough, my dad taught me how to drive a stick. I drove the jeep daily to high school. I'd later drive a 5-speed manual Crash truck when I was stationed in Guam.

One day, coming home from school during a downpour, I traveled down a muddy road to our house. Long story short, the jeep slipped sideways and hit a pothole. It flipped over onto its top—thank God there was a hard top. I got banged up a bit, but after recovering, I cut the top off and drove it around. But I digress…

While in Bremerton, I figured it was time to buy my own vehicle. My dad had sold the jeep after I left for the Navy, and since I was also working part-time at a gas station, I had a little extra spending money. Besides, I made Second Class Petty Officer (ABH2) while in the shipyards and was making $492 a month.

I ordered (you could order cars back then and get what you want) a new 1976 Chevy pickup truck from a Gig Harbor, WA dealership. "3-on-a-tree," 350 cu. in. engine, and I put a camper top on it. Total cost was about $4,500. I called her the "Chet Mobile."

Everybody remembers their first car or truck they bought, right? How many of you had a nickname for yours?

Olympic Community College

One of the other things I wanted to accomplish while in Bremerton was to take a course or two at the local community college. I mean… why not? I certainly had the extra time and didn't frequent the bars as much.

I enrolled in a basic electronics class at Olympic Community College. Prior to this class, about the only thing I knew about electricity was how to turn a light switch on and off… and even then, only when I was sober. But I digress…

Since I knew my 4-year military commitment would end in July '77, I was trying to figure out what I wanted to do with the rest of my life.

At that time, the class cost $83. We met four evenings a week. The Navy picked up 75% of the tuition, so I only had to lay out about twenty bucks. Pretty good—I was learning how to work the Navy system.

Taking this course set me on another "journey" in my life. I would eventually return to college (University of Florida) and earn a Bachelor's degree in Electronics Engineering. A few years later, I'd graduate from UF again with a Master's in Engineering. That set the foundation for my career. But I digress…

Goodbye, Bremerton

After January 1, 1977, every letter I wrote home I began with the "188 days and a wake up" or however many days I had left in the Navy. Am I the only one that did that?

I was getting short and I could feel it. I tried not to let it affect me, but as many military men and women will confess, your attitude changes. And yes, eventually, I did make my own short-timer's chain, though the V-1 Division MCPO didn't like to see such things.

Master Chief Breig (V-1 Division MCPO) pulled me aside a few times to talk about reenlisting. While he and I *rarely* saw eye-to-eye, I *politely* listened. In 1977, the Navy offered ABHs $1,200 to re-enlist for 6 years. I'd also heard they offered "Nukes" (nuclear submariners) $20,000 to re-up for 4 years. That was an easy, "Thanks, but no thanks" for me.

While a few shipmates may not have appreciated our time in the Bremerton shipyards, I tried to take advantage of both the area and our time off. The shipyard experience is *no comparison* to a cruise. *Nothing* will ever match the challenges and excitement of working on the flight deck of an aircraft carrier.

However, this was a lesson that stayed with me. Some jobs aren't as glamourous as others, but you do it because it's your job, like it or not. And sometimes the location of those crappy jobs far outweigh the dullness—take advantage of these opportunities when you can. And yes, I still do that.

I knew the Navy was not for me, but I would soon complete my four-year commitment. During my time in Bremerton, I felt my attitude changed and I began to wind down my short ***Three Years, Eleven Months, & 29 Days*** Navy career.

Chance.... or Fate?

Had I crossed over to another ship and made another Westpac cruise instead of going with *Kitty Hawk* to Bremerton, I would have missed out on my first snow experience and purchasing my Chevy truck. Most important, I may not have discovered my love for math, physics, and electronics that led me to become an Electronics Engineer.

Decisions & choices—I'm at that point in my life when I look back on some of these paths I've taken and wonder if they were really *my* decisions or God's. I would ask myself that very question many more times.

(43) Port Call: Hong Kong

A quick word…

I have many great memories of my time in the service and share them in this weekly blog, "A Veteran's Journey." I blog twice a week: Tuesday morning is the "main journey" and Friday afternoon I blog a "Navy Friday Funny."

Most of my blogs are about the Navy, but every month or so, I'll blog about my dad, who served with the 80th Infantry Division, and my father-in-law, who served in the 82d Airborne and parachuted into Normandy on D-Day, both in the ETO during World War II.

I share these blogs on Facebook, Twitter, Instagram, LinkedIn, and Pinterest.

Reconnecting… and memories

Over the past eight months (I started blogging in November 2019), I've heard from many sailors who have been stationed in various locations around the world. I've even "found" a few former shipmates.

It's always great to reconnect with shipmates I knew and served with. I also know that through these blogs other shipmates have also connected up. It often brings back great memories to share together, and one shipmate may remember something you'd forgotten.

Funny how that works: different viewpoints.

Hopefully my blogs may trigger a pleasant memory, but I know other times they may hit a sore spot. I hope that doesn't happen often. But we all faced many of the same obstacles, challenges, and enjoyed many traditions. I try to focus on the more positive experiences, though periodically I find it necessary to mention events that went awry. It's a fact of being a military veteran.

In these challenging times of COVID, this is one way for me to keep busy. I know we're all facing daily struggles we've never seen before, there's no doubt about that. So I hope these "journeys" help bring back some good times you experienced during your time in the Navy.

We are all in different phases of our lives, but we all have something in common: we served in the United States Navy and to me, I'm proud of my military service. I hope you are, too.

# # #

And now… on to Hong Kong

Kitty Hawk's Westpac '75 cruise took us first to Hawaii for a short stopover, then for a two-week transit to the Philippines for our overseas forward deployment. We'd be in and out of PI from May 1975 until we left to come back to the states in December 1975.

From PI, we cruised to two additional overseas ports: Hong Kong and Japan. Even though there was scuttlebutt on the decks about heading southwest to Christchurch, New Zealand (and crossing the equator), that was a no joye.

Two things stick out for me from our Hong Kong visit: one, a tattoo which I blogged about earlier. The second, "slightly" more embarrassing event, was that while I was in control of the B&A crane (Boat & Aircraft), I sent a 1,500 pound hoist through the windshield of the Admiral's barge.

That was NOT a good day. But I digress…

We left Subic Bay, PI on August 7, headed northwest for Hong

Kong. At that time, it was still under British rule and we would be there for a week. China was closed to Americans, but I "heard" several shipmates who "found a way" to cross the border for a day to explore.

Dress White's Inspection

When we left PI, it rained for four days straight. I'm not talking a little "Seattle-type" drizzle. I'm talking a continuous, big 'ol frog-strangler torrential rain like we have in Florida. It was pretty miserable on the flight deck, even though we all wore rain gear.

We were scheduled for a "white's inspection" on the flight deck, too. That made a lot of standard Navy issue sense—forty-plus knot winds with rainstorms.

Heck, we *never* wore our white uniforms… they were stuffed away in a locker somewhere onboard. Well… we did when we were on Shore Patrol duty, which only happened a few times (for me) on Westpac '75. And oh yeah, when we manned the rails. So I guess there were times when we actually broke out the white uniforms.

But because of the inclement weather, the Old Man *finally* decided that instead of having a ship-wide white's inspection, each division would hold their own on the hangar deck. Well, hats off to the skipper for that call.

We arrived in Hong Kong on August 13. We didn't have a pier to pull up alongside like we did in PI, Japan, or Hawaii. Instead, we anchored in the harbor near Green Island and took ferries to get to the port.

We also had "max liberty," meaning we only needed to be on the ship if we had duty. Several shipmates' wives had flown to Hong Kong to spend time with their husbands. We didn't see them all week, but when we did, they all had a big ol' grin. But I digress…

It turned out that I had duty the first day in port… bummer. But I made a few extra bucks from some guys who didn't have enough cash: "Lend $20, payback $40 on payday."

The Admiral's Barge

I was one of a few in Crash who knew how to operate the Boat and Aircraft (B&A) Crane, which was a *huge* crane that swung out from the starboard side of the ship over the water and could raise and lower either aircraft or boats.

The Admiral was on board *Kitty Hawk* and wanted his barge (boat) in the water. "No problem," I said. The Admiral's barge was kept aft on the hangar deck, so the V-3 Division Yellow and Blue Shirts towed it to Elevator #3, which had been lowered to the hangar deck level. Bos'n Robuck, the Crash Boss, was with me to make sure everything went smoothly.

I lowered the B&A Crane hoist so the guys on the elevator could hook up the barge to the massive 1,500 pound pulley. There were three Boatswain's Mates in the barge to ensure sure everything was hooked up properly. They used a heavy shackle to connect the four straps to the hook on the pulley. Nothing major, this was standard operating procedure. Seas in the harbor were four to five feet, but *Kitty Hawk* was so heavy and so big, it didn't sway much.

I operated the crane to lift the barge and the three BMs slowly off the elevator, then slowly swung the crane so that the barge was over the water. Easy, peasy. Next, I slowly lowered the barge until it was in the water—all SOP.

While it was *not* rough seas for the carrier, it was pretty rough for those guys in the Admiral's barge. They struggled to wrestle with a 1,500 pound pulley in four to five foot waves.

It wasn't easy. Every time they signaled me to lower the hook so they could undo the shackle, they got bounced around. Finally, one last time and they were able to undo the shackle. But at that particular moment, the barge lunged backwards and the 1,500 pound pulley went right through the windshield of the Admiral's barge.

It was not a good day for them or me.

We all worked hard trying to prevent such mishaps, but with the rough seas, it was next to impossible. I was worried I might get written up for that incident, but I heard nothing more from it. In fact, Bos'n Robuck told me I did an outstanding job, despite the high seas. I guess if he was OK, I was OK.

In the military, you periodically get "atta-boys," meaning a job well done. But in the same sense, one "oh crap" will cost you about

25 atta-boys. Fortunately, I didn't lose any atta-boys that day.

The Admiral had his barge, albeit a little wacked up, but I'm sure the BMs were able to fix the windshield.

Finally… Liberty

I got liberty the second day in port, so I headed out with a few friends. The exchange rate in 1975 was five Hong Kong dollars for one US dollar. Steve Cummings, who had just been promoted to a third class petty officer and went from a Blue Shirt to a Yellow Shirt, Smitty (Jimmy Smith), and I decided we would find a tour and see the sights. We knew we could always grab some beers later at the China Club, a popular hangout for sailors.

We found an all-day tour that took us up to Victoria Park, the highest mountain in Hong Kong, to have a great overlook of the harbor. *Kitty Hawk* was certainly the biggest ship in the harbor, and she was just outright beautiful. We all felt proud to be a part of her crew.

We also spent a little time at Tiger Balm Garden, which had a seven-story pagoda. It was built by Aw Boon Haw and his family in 1935 for a cost of $16 million Hong Kong dollars.

Part of the package included a boat tour through Aberdeen, the floating city in Hong Kong. Here, entire families lived on small boats called sampans, usually tied up parallel to a pier. That means if you wanted to get to the boat on the outside, you had to step onto the other boats between it and the pier. I've never seen such a crowded place on the water. I don't know how they did it.

We ended up at the Tai Pak Restaurant, a floating restaurant with exquisite decor. They served us a full six-course meal that was absolutely fantastic. I had learned earlier in the Philippines *not* to ask what I was eating, especially if it tasted good. I didn't want any surprises.

Because this was a floating restaurant, water cost three Hong Kong dollars, which was equivalent to about sixty cents in US dollars. Beer was free, and we could drink all we wanted. I really enjoyed that place. Lots of great food spread out over a long period, and all of us on the tour had a great time.

Cost of Living "Comparison"

I wrote my parents and told them Hong Kong was a wonderful place and that I was actually going on tours instead of hitting the bars. That was something I never did while in the Philippines (tours, that is; I certainly frequented the bars).

I started comparing the cost of beer: $1.50 US a bottle in Guam, 40 pisos ($.45 US) in PI, and $4 HK ($.85 US) in Hong Kong. It was a crude currency comparison, but I think they got the picture. The same item in different countries cost different amounts.

It is amazing how the world economy works—just compare the cost of beer and you'll have a good idea of the cost of living. Am I the only one that did that?

Oh yeah. One thing I forgot to mention about the ferries to/from (specifically, BACK TO THE SHIP)…

The trip from *Kitty Hawk* to the port was during the day. The trip back, at least for me, was usually *after* a full day of touristing, one or two (*or more*) drinks at the China Club. Some nights were a little later than others.

But every trip back to *Kitty Hawk* from the port was, how shall I say? "rougher" than the trip in. You'd think one would learn a lesson here, but since someone else was driving the boat, it really didn't matter.

Tell me I'm not the only one who experienced this? Go ahead… I'll wait.

Back in PI

We returned to Subic Bay on September 4 for a much needed rest. We had almost two weeks in port, though we had a normal 0730 to 1530 (7:30 to 3:30) work schedule. I took advantage of the "down time" aboard ship, even taking duty for several other shipmates.

I thoroughly enjoy my six days in Hong Kong, primarily because it was a new country with new traditions and the fact that I took time to explore the area.

(44) When Life Hands You Lemons…

Let's face it, we've all been handed a lemon or two during our lifetime. And I'm not just talking about the fruit. While I'd heard (& used) this expression for years, I had to research its origination.

Professor Google reports it was first used by Elbert Hubbard in 1915: "He picked up the lemons that Fate had sent him and started a lemonade stand." But many people attribute the expression to Dale Carnegie in 1948: "If You Have a Lemon, Make Lemonade."

The theory behind "making lemonade" is to encourage optimism and a positive, "can-do" attitude in the face of adversity or misfortune.

Being a Navy veteran, we have a different term we use to describe trying to make something unpleasant seem more appealing than it really is, but it often becomes a futile effort. Before I get to the "Navy term," let me share a few times when I felt the need to "*polish* a lemon."

Making Lemonade in Guam

My first duty station after ABH "A" school was NAS Agana, Guam. Not that Guam handed me many lemons, but there were times when things could've been better.

My job in Guam was simple: wait for aircraft to crash, then put out the fire and rescue people. Hence the name of our division, "Crash & Rescue."

When I first arrived, our shift change was at 11:30 a.m. That only allowed us a half day of fun in the sun. We petitioned (or whatever you call it in the Navy) to have the shift change at 0800, which was much better all-around for everyone. The Crash Commanding Officer, Bos'n Joye, approved our request, giving us a full day of sun on our time off.

Driving Crash Trucks

Our normal workday in the Crash Barn was 0800-1600; after that, we had "holiday routine." We spent a lot of time training, both in the class room as well as hot drills (simulated fires).

But every 24-hour shift, we were assigned to one of five Crash firefighting trucks. Three of them spent two 4-hour shifts out on what we called the "alert spot" or "hot spot." That was a specific area between the two parallel runways, about midfield. This would be the first truck on the scene in the event of an aircraft crash.

In Guam, the runways served both the Navy and the civilian airport—we had to be prepared for both.

There wasn't a lot to do out on the hot spot and thank the good Lord, we didn't have any crashes during my time on the island. We had plenty of aircraft emergencies, though. That's why we always needed to stay alert, even in the middle of the night. I wrote about "That Adrenaline Rush" earlier.

It wasn't the most glamorous job—sitting in a fire truck out in the blistering sun for 4 hours—but we made do.

However, the driver had the best job of the 4 or 5 crew members on the truck. I decided early on to "make lemonade" and learn to drive the Crash trucks—all of them. It wasn't long before I was certified to drive the three different types of Crash trucks plus the 500-gallon tanker (that was a beast, by the way).

Private Pilot? Why Not

I was only 18 years old when I arrived in Guam in November 1973. After a few months getting accustomed to my new surroundings, I grew tired of frequenting the bars and throwing money away on booze. It was time to find something else to do in my off-duty time.

Near the Crash Barn was a flying club, complete with several privately-owned planes (Cessnas & Apaches). Bruce Hallowell, a fellow Crash crew member, took lessons there and talked a lot about flying around the island. I thought, "Why not?"

After my first $9 "intro" lesson, I was hooked. I learned to fly while stationed in Guam and earned a private pilot license. Costs? In 1974, an hour with a flying instructor, including the plane rental, was $18… total!

The island of Guam isn't that big: 4-8 miles wide by 32 miles long. There were a few "restricted" areas where you might be shot down if you flew too close, but those were clearly marked on the map. Besides, there were plenty of other areas where I could fly. Talk about clear, blue water and sandy beaches… Wow! I even took a few buddies on flights around the island.

When you're in the Navy, and I'm sure in almost every branch of military service, it doesn't take you long to learn "the Navy way," and how to work the system. That knowledge of "working the system" would come in handy throughout my life.

Making Lemonade on the Carrier

I had to think about this one for a bit, but there were those times (you know what I'm talking about) when life just didn't seem all that fun on the flight deck of *Kitty Hawk*. That didn't happen often, but when it did, I think we all figured out how to "make lemonade."

I was an ABH (Aviation Boatswain's Mate-Handler) and worked on the flight deck in V-1 Division, originally assigned to Crash. Later, after the Bremerton Shipyards, I became a Yellow Shirt director. So my viewpoint may be a little different from others who served in the Navy.

I never dove headfirst into grease, like the ABEs in the V-2 Division: they're the guys who worked the Launch Catapults, the Jet Blast Deflectors, and the Arresting Gear. I don't think a day went by that they weren't covered in grease.

And I never swam in JP-4 and JP-5 jet fuel, like the ABFs in the V-4 Division: they're the guys who kept the aircraft fueled day in and day out. V-3 Division (also ABHs) worked the hangar deck, and I'm sure they had their "fun times," too.

V-5? I'm sure they did something <grin>… but the only time I saw them on deck was when they were escorting the Miss America Tour to and from the COD (Carrier on-board Delivery) during Westpac '75. I'm kidding, of course, because we all know they played an important role in air traffic control. They just *always* seemed to stay clean.

FOD Walk Downs…
… or how to keep from having to make lemonade

FOD (Foreign Object Damage) walk downs weren't always fun, but we all knew they were necessary. The squadron maintenance folks certainly appreciated what we did.

For those of you who've never heard of a FOD walk down, it's basically a gathering of every able-bodied flight deck member. Before flight operations began, we lined up across the bow, shoulder-to-shoulder, and walk aft down the flight deck toward the stern.

The goal was to look for any object (FOD) that could be sucked up into a jet intake, thus potentially damaging the engine and ruining a perfectly good flying machine. Even a small wing nut can tear up an engine.

It was also a way to meet up with other shipmates either in another division or squadron and chew the fat. A chance to catch up, so to speak. Or, in some cases, an opportunity to collect the money you won the night before playing an all-night poker game. But I digress…

Making Lemonade in the Shipyards

I wrote about life in the Bremerton Shipyards in two earlier blogs. While we worked a "normal" workday, 0730-1700, there wasn't a lot to do at night in Bremerton.

I also knew that the Navy would not be my lifelong career after my 4-year commitment.

My discharge date was July 8, 1977. I "felt" my attitude change during our time in the shipyards ('76-77), I made a conscious decision to take a basic electronics course at the local college, Olympic Community College. That course set my journey for my "professional" career(s).

Several of us in V-1 had studied and took the second-class petty officer exam in August while on Westpac '75. My good friends, Glenn Law, Ed Boes, Gary Borne, and I spent a lot of time studying together during the cruise and it paid off… sort of. While we passed the exam, we were all "PNAs" (Passed, but Not Accepted). We didn't have enough time in service.

The four of us also took the second class exam again in February. At the time, one of my duties was the V-1 Division Training Officer (that's was nice "skate job" while in the shipyards), so there was no excuse to not to study for it.

In late April '76, I learned that I had passed the E-5 exam and would be promoted to a second class petty officer, ABH2, on June 16, 1976. My salary would now be a whopping $492 a month.

Normally, an E-5 is required to have three years of service before advancement, and I was one month short (I joined in July 1973). But because I had joined the service as an E-2 (Airman Apprentice) with a year of college under my belt (even though I flunked out of college after my first year), the Navy gave me the waiver.

I guess that year at FSU (Florida State University) was worth it, even though they asked me to leave.

Current COVID—Making Lemonade

I've never been shot at in a firefight or under an artillery barrage like some of my WWII and Vietnam veteran friends. I can't imagine…

But like many of my flight deck shipmates, I've been in split-second life & death situations multiple times. And… a few more while I was a City of Gainesville firefighter after the Navy.

This current COVID crisis, or whatever you want to call it, got me to thinking about the "lemonade" I've been making since March. We all have our ways of dealing with this situation, but for me, I've reached back into my old seabag and retrieved some of my Navy traditions, including how to "polish *this* turd."

It won't be this way forever, but I'm assuming it will be around for a while and I'm taking my own precautions to remain safe and healthy. I've discovered my "new normal" isn't really all that different from my old normal.

Of course, I miss spending time with family & friends, but we've also learned to adapt to alternative methods.

My wife and I are both retired, so we don't feel much the strain and stress many other folks experience. I do *not* take this lightly; I try to help when & where I can, but there's only so much a fella can do.

Video Chatting

As an example of my new normal, my 8-year-old grandson in California calls me several times a week on a video chat app and we play solitaire (with cards), games, and he shows me things he's building, and reads stories he's writing (he's a writer, too <<wide grin>>). Now and then, his 6-year-old brother joins in on the video chat and plays games.

I know they'd rather be with their friends and in their summer activities. But for now, spending a little time with their granddad is, to me, precious. I hope I'm able to help them pass a little time during the day and explore some new ideas from their ol' Navy granddad.

Fresh Air & Exercise Do Wonders

Another example: my wife and I are exercising more; well, I am—she *always* has. We take long (10-25 mile) bike rides around the city once or twice a week, we're both into Yoga, we walk and swim.

The "Y" is open, but I'm not ready to go inside to lift weights just yet. We do, however, use their outdoor swimming pool (with reservations), and we swim laps (we're over ½ mile these days).

All of these outside adventures help us clear our minds of the daily onslaught of [insert your own term here].

Writing Keeps Me Busy

I'm also writing more. I started this weekly blog (A Veteran's Journey) last November. I finished writing a book, NEVER FORGET, that's currently with an editor. Who knows where it will end up, but I certainly enjoy writing and as I tell my wife, it "keeps me off the streets."

I also blogged a three-part series earlier, "Writing Your Own Memoir." Just in case you might want to write about *your* time in the Navy (or Army, Marine Corps, Air Force, or Coast Guard), feel free to download the entire 3-part series directly from my website.

It may provide you some guidance (and a distraction) over the next few months. I can guarantee you that writing a short memoir or jotting down your "adventures" will make an impact on both you and your family, *and* generations to come.

I also have another couple of books I plan to write—but I need to publish NEVER FORGET first (wish me luck!).

When I look back on my *short* 4-year time in service, the Navy definitely helped me to learn how to "accept the things I cannot change, the courage to change the things I can, and the wisdom to know the difference."

Or in laymen's terms… "When life hands you lemons, make lemonade."

(45) WWII Baptism by Fire, 76 Years Ago

My dad, Capt. A.Z. Adkins, Jr., fought with Patton's Third Army during World War II. "Pop" was a 22-year-old officer, an 81mm mortar section leader in Company H, a heavy weapons company in the 2d Battalion of the 317th Infantry Regiment of the 80th Division.

Originally, 80th Division WWI men were recruited from Virginia, West Virginia, and Pennsylvania, all states in the Blue Ridge Mountains, hence the nickname, the "Blue Ridge Division." The division motto was, "Always Move Forward."

The 80th Division crossed the Atlantic Ocean on the *Queen*

Mary, leaving the US on July 4, 1944. Almost 28,000 troops, including other divisions, arrived in Greenock, Firth of Clyde, Scotland a few days later on July 7, greeted by a performing group of bagpipes.

I cannot imagine what my dad experienced during the war—I was Navy ('73-77) and assigned to NAS Agana, Guam for 15 months, then to *USS Kitty Hawk*, an aircraft carrier, for 2½ years. No one ever shot at me and I never experience an incoming artillery or mortar barrage.

However, I have WWII and 'Nam friends who have experienced those circumstances, and let me tell you, it's something they rarely talk about, but it's also something they'll never forget.

The following excerpt is from my WWII book, ***You Can't Get Much Closer Than This: Combat with the 80th 'Blue Ridge' Division in World War II Europe,*** published by Casemate Publishers in 2005. This is the story of my father's "Baptism of Fire."

# # #

Normandy landing @ Utah Beach

In late July, the division prepared for the trip across the English Channel. The Normandy beachhead had grown substantially since D-Day. We boarded trains at night, pulled the curtains down over the windows and eventually arrived at Southampton, England, where we boarded ships for the 21-hour channel crossing.

When we got close to the beach, we climbed over the ship's rail and down a cargo net onto a landing craft. We all huddled down as we headed over the last few hundred yards to the beach. When the boat grounded on the beach, the ramp lowered and we waded in water above our knees to cover the short distance to the beach.

I stepped ashore at Utah Beach on August 5, 1944, at 3:00 in the afternoon. It was D-Day plus 60.

Assembling an entire division

We reached the regimental assembly area, about 500 yards from the beach, and climbed aboard trucks for a ride to our bivouac area near Vareville, France. The assembly area had a peculiar and unforgettable odor: dead men who had begun to decay. You never forget the stench of a decaying corpse.

We inspected and cleaned our weapons while we waited for the rest of the division to arrive. We were tense, but ready. On August 6, we entrucked again to the vicinity of Coigny, France.

We were close to combat, but not yet in battle. Everybody was on the alert. Any fast-moving jeep, command car, or truck meant the division was moving out. The unexpected appearance of a high-ranking official meant the same thing. Rumors flew all over the place and I'll admit, I found myself caught up in the chaos.

On August 6, the Germans counterattacked from Mortain, France. Their aim was to slice the Third Army supply lines near Avranches and cut off part of Patton's troops from the rest of the Allied forces. The 80th Infantry Division received its battle orders the very next day: help stop the German counterattack.

First taste of war

Two days later, we moved by trucks forty miles to Pain D' Vaine, on to LeMans, and then to the vicinity of St. Hilaire, which was literally in shambles. Little shops and houses were smashed and gutted, their walls stood jagged and roofless. The roads were shell pocked from artillery and littered with battle debris. Charred enemy vehicles lined the route. This was our first glimpse at what artillery can do in a war.

We reached the assembly area north of Evron, France, and received orders at 7:00 p.m. to attack and seize Evron. The 1st and 3d battalions attacked in echelon, with my battalion held in reserve. The objective of this attack was to seize Evron and gain the high ground to the northeast. The 317th occupied the town by 10:00 a.m. the next morning. This was our first taste of battle. Even though I wasn't in the middle of it, I heard from those who were.

Argentan, France

The next few days, we marched and entrucked through several small towns with minimal resistance. In mid-August, we moved into an assembly area near Argentan, the town through which the German Army intended to escape to the east from the Falaise Pocket. This pocket was caused by the closing up of the Canadian forces coming from the north and the American forces coming from the south. This was to be our first major encounter with the Germans and our "Baptism by fire."

On the morning of August 20, the 3d Battalion attacked Argentan, contacted Canadian units, and sealed the fate of the German Seventh Army, which had been completely smashed in the closing of the Argentan-Falaise Gap. Ten thousand Germans died, and another 50,000 captured. After Eisenhower visited the site, he wrote: "It was literary possible to walk for hundreds of yards at a time, stepping on nothing but dead bodies." Falaise became known as "The Killing Grounds."

Heroism

During the attack, Pfc Hoyt Rowell, a medic assigned to the 317th Infantry, was rendering aid to wounded riflemen under a hail of shells and bullets. When friendly artillery began to fall nearby, he raced across an open field to an artillery observer and lifted the fire. Returning across the same hazardous route, he continued to aid the wounded. For this display of heroism, Rowell earned the first Distinguished Service Cross (DSC) awarded to a member of the 80th Division in the war.

This was our first real test in battle. Our orders were to seize the high ground northeast of Argentan, then swing southwest to take the city. We practiced what we trained, but it was much different. No one at Camp Laguna (Arizona) had shot back at us, let alone dropped mortar shells all around us. Adrenaline kicked in and affected men in different ways. Some seemed exhilarated while others fell apart from fear. A few men became "shell-shocked."

Taking our objective, one at a time

My battalion attacked straight up the south slope of Hill 244, our objective. This high ground controlled Argentan from the north and the Argentan-Trun highway in the Foret de Gouffern.

Enemy resistance was strong; machine gun fire slashed into Companies E and G, which suffered seventeen casualties. One was Pfc Earl Goins, Company E, a North Carolinian, who earned the DSC when he rushed ahead through deadly fire and silenced a machine gun blocking the advance of his company. Goins's heroism allowed his fellow GIs to outflank the remaining enemy gunners, who eventually withdrew. By noon, both companies were atop Hill 244.

We take Argentan

There was confusion everywhere. American and British warplanes roamed the skies and bombed and strafed German tanks, trucks, and infantry columns. Hundreds of German soldiers dropped their weapons and raised their hands, looking to surrender to the nearest Allied forces. Others seemed too afraid to fight or flee and huddled near their stalled vehicles.

By nightfall on August 20, the 80th Division had taken Argentan. In the three-day battle, the 80th destroyed fourteen enemy tanks and captured a thousand German soldiers. We captured or destroyed vast quantities of German equipment and supplies, including a dump with 27,000 tons of ammunition and a map depot. But our victory was did not come easily. The 80th Division lost more than 400 soldiers, killed or wounded.

My battalion had started cross-country at dawn one morning and was still going at midnight. We were so tired some men were falling to the side of the road and had to be prodded back in line. A little after midnight we saw a big fire ahead. As we got closer, we could tell that a whole village was burning. We didn't see a living thing as we entered the town.

Suddenly, an old woman came out of the rubble and in a frightened voice asked, "American, American?" Someone said, "Yes, we're American." The woman was silent for a moment and then as if she suddenly realized that we were Americans and not Germans, cried into the burning rubble, "Les Americans ici, Les Americans, ici!"

Dear Mom & Dad

In a letter to my mom and dad, dated Thursday, August 24, 1944, I wrote:

"Mom & Dad, I have been doing a little fighting lately, and have been in one major battle. You probably read about it in the papers. I came through it without a scratch.

I have been 'baptized' to fire, so now I know all the tricks. Therefore, there is absolutely nothing to worry about. I am a good soldier, love and believe in my God and my family and therefore I will be home before very long.

All you have to do is pray for me and everything will be alright."

# # #

My dad, my hero…

My father was wounded twice during the war and earned the Bronze Star Medal (Valor) for action in early October 1944 at a place called Sivry, France. I'll blog about that later.

After the war in Europe ended on May 8, 1945, he was part of the "Occupational Force" and remained in Austria, Czechoslovakia, and Germany until he returned to the states in January 1946.

My father attended law school at the University of Florida here in Gainesville, met my mother & married and had two kids (me & my older sister, Anne).

Pop rarely talked of the war and even when he did, it was usually about a good Army buddy or a humorous story, often repeated. He *never* talked about the "bad stuff."

Then for Christmas 1985, he gave me and my sister, Anne, a typed 200-page, 3-ring binder titled, "**A World War II Diary**." For the first time in my life, I read my father's WWII story (again and again) and realized the horrors he not only experienced, but those he also witnessed.

I can't for the life of me imagine how these courageous men carried those memories with them throughout the rest of their lives. But my dad, as most WWII veterans, was always proud of his service to his country, his community, and the love of his family and God. I believe he always continued to follow the 80th Division's motto, "Always Move Forward."

We lost Pop in June 1989 of cancer, three days before his 69th birthday. He was my rock, my friend, and my hero. To this day, he continues to inspire me, especially in these challenging times.

(46) The "Oh Crap" Bag… What in Yours?

We're right smack dab in the middle of hurricane season here in Florida, which is "officially" designated between June 1 and November 30. I've lived in Florida most of my life, as has my wife—she grew up in Miami; I grew up here in Gainesville. We've been through many hurricanes and I can tell you from experience, no two are the same. You'd think we might get used to them, especially as we get older, but I've got news for you… you don't.

I never heard the term, "The Oh crap bag" when I was in the Navy. It was something we just didn't need to worry about. When you're a young, unmarried sailor stationed in NAS Agana, Guam or aboard *Kitty Hawk*, you don't have that much to haul around. Simply put, if it didn't fit in your seabag, you didn't need it.

I first heard the term when I worked as a firefighter with the City of Gainesville, not long after I left the Navy. During a call to a house fire, one of the more "seasoned" firefighters made the comment, "I sure hope they took their 'Oh sh*t bag' with them." After the fire was out, which BTW, devastated most of the house, I caught up with him and asked what he meant.

That's when I learned that the "Oh crap bag" is a prepacked bag with important stuff you grab on your way out the door when you have to leave in a hurry and you're not sure if your house will still be there when you return. It takes the guesswork out of what you need when you need it.

It's not something to joke about… this is real.

Florida hurricane season

Hurricanes come ashore somewhere in Florida almost every year. Some are worse than others. But in every case, we have the "opportunity" to watch the Weather Channel and the local news to see where the storm is, where it's expected to hit, and decide from then, whether or not we need to leave the house. In other words, we usually have a lot of warning (and I might add, a lot of *anxiety*) ahead of time… that is, for hurricanes.

At the beginning of the season *and* when a hurricane approaches, the newspapers, local TV stations, and social media *always* provide a "checklist," commonly called a "Disaster Supply Kit." This contains common sense supplies, such as a flashlight & radio (with batteries), first aid, and pet care items. This year, they've included COVID-19 items like face masks, hand sanitizer, and cleaning wipes—like that's one more thing we need to worry about. Sheesh!

Hurricanes are *NOT* the only worry…

But—you may not know this—hurricanes can spin off tornadoes. That's a totally different story. You *never* know when or where *those* disasters hit. And unlike hurricanes, there is rarely any warning.

My wife and I lived in Idabel, OK for a year in the mid-80s. I worked as an electronics engineer for Weyerhaeuser Company, a forest products company. We actually started *this* "journey" in 1982 in Federal Way, WA, midway between Seattle and Tacoma in the Pacific Northwest… beautiful country. We were there for a year. I blogged a little about that earlier.

However, I musta stepped on someone's toes, though, because after my first year, the company transferred us to a pulp & paper mill in Valliant, OK, located in the armpit of southeast Oklahoma. I say that jokingly. It was a lovely place, beautiful country, we made some good friends, and… our daughter was born there.

But we were 10 miles from Texas and 20 miles from Arkansas. If we went out to dinner in Idabel (that's where we lived), we'd eat at a Sonic drive-in restaurant or at Bob's Catfish house. Not a lot of night life in Idabel.

But I digress….

We were fortunate enough to NOT see or experience any tornadoes while living in what is commonly called, "Tornado Alley." But we had an "Oh crap bag" ready, just in case.

I've been through a tornado or two here in Gainesville, both as a firefighter (that was interesting) as well as a citizen (that was challenging). They are scary and—my wife and I have talked about this several times—even if we needed to evacuate, where would we go? In Oklahoma, people had underground storm cellars; in Florida, if you dig down more than two feet, you hit water. Not a lot of options here other than to hunker down.

And yes, we've spent more than one night sleeping on the floor in the hallway, away from windows and doors. That's not a lot of fun, trying to get some sleep but staying mostly awake, listening to the wind howl and the trees sway, hoping that nothing comes crashing down through the roof of the house.

No power for days

There was that time, though, a couple of years ago when a hurricane came right up the middle of Florida. While we didn't sustain any house damage, it blew ("micro bursts") several huge oak trees down in various parts of our neighborhood, which took out the overhead power lines. Long story short, we were without power for six days. Fortunately, we had running water and our propane gas

BBQ grill, so at least we could make coffee and cook a few meals. We kept our cold stuff in ice chests and coolers.

The first few days were okay, and we could sleep with the windows open because it was cool at night. But then several neighbors cranked up their diesel generators. They kept them running day *and* night. Not that it reminded me of my time on the flight deck behind a jet exhaust, but I found myself more than once reminiscing those times, "triggered" by the smell of diesel fuel *and* exhaust.

I learned something new, though. The Gainesville house we bought when we moved back from West Virginia in 2015 had been completely renovated including new plumbing, with the new water pipes run through the attic. Since it got hot in the afternoons, that meant the water in the pipes was also hot (we had an electric hot water heater). So back to the standard Navy-issue "three-minute" showers.

After six days, the utility crews finally restored power. It turns out that on our neighborhood block, two power poles had snapped in half during the storm. There were hundreds that had snapped city-wide, and the utility company simply ran out of replacement utility poles. But we were thankful that the damage wasn't any worse.

Losing a house

I have a close friend who lived on the St. John's River in St. Augustine. Over the years, she & her husband have more than once hunkered down during hurricanes. They were far enough inland that the hurricanes didn't affect them as if they had lived directly on the coast. That was… until a few years ago.

Their house was about an hour and a half from us. In the past, she'd come over to stay with friends here in Gainesville while her husband remained behind to ride out the storm and "take care of anything" he could during the storm. While they'd had damage to their wooden dock a time or two during past hurricanes, *nothing* compared to the unexpected storm surge caused by this particular hurricane two years ago.

Her husband recorded a dozen short videos on his phone from inside the house as the water began to rise in the river—the "storm surge." First it was two feet below the dock, then an hour later, water was even with the dock. Then the dock disappeared, but the water still had not made it over the river bulkhead.

Not much later, the waves were strong enough to crash over the bulkhead. The house was about 20 feet away from the river. Then the river peaked over the bulkhead, but not in the house (yet). Then, the waves began to crash up against the sliding glass doors knocking them off their tracks, but he was able to put the doors back in and physically hold them in place, keeping the water from entering the house.

A little while later, the water was not only *under* the house, but coming in through the front door. It wasn't pretty. By the time I got over there the next day, the water had subsided, but he was mopping it up the best he could. Water was still in the garage, about a foot deep, but he had managed to save as much as he could. We pulled the carpet, mopped and squeegeed, and got most of the water up off the floor, but there was a distinct waterline about six inches up the walls.

I know they weren't the only ones whose home was damaged that day; many neighbors also experienced damage. The thing is, that house had been on the river for forty-plus years and had seen many hurricanes. This one, though, was the one that did the most damage; so much, that the house couldn't be saved.

They are safe, though, and found another place not too far from their old house. It's near, but not as close, to the St. John's River.

Mother Nature doesn't kid around

We live in Gainesville, Florida, located a couple of hours from each coast, about an hour south of the FL/GA line, and two hours north of Dizzy World. We don't get the flooding and winds that the coastline communities receive. Yes, we get damage now and then, but mostly, it's debris from the trees. After a major storm, we usually fill up four or five garbage cans of this "salad" debris.

We're at that age where we don't want bad things to happen to anyone. In one sense, you're happy that the hurricane didn't directly affect you. But on the other hand, you don't feel right hoping it goes somewhere else, because you know it's going to hit somewhere and wherever that somewhere is, it ain't gonna be pretty.

Survivor's guilt

I'm sure I'm not the only one that has this "selfish" thought. While I'm talking about hurricanes in this post, I know also from experience on both the flight deck and as a city firefighter, that no matter how careful you are, sometimes, people get hurt. But if you're "lucky" enough to *not* be hurt, you sometimes still carry a feeling of guilt.

Over the years, I've spoken with many veterans, including WWII, Vietnam & Vietnam-era, and GWOT (Global War on Terror). Many of us periodically experience a "trigger" that takes us back to a not-so-good memory of an event that occurred during our time in the service. We rarely like to talk about it, especially with our families and our kids. One thing I have noticed, though, is that it's often easier to talk with another veteran who's shared a similar experience—someone who's "been there and done that."

The "1000-yard stare" is a common term, I'm sure you've heard. Most of my veteran friends know what I'm talking about. I experience that periodically. It's not necessarily a bad thing to remember, because it helps remind me of the good times I've had in my life, too. Sometimes, I wish I could go back in time and "tweak" a few things, but that's not possible.

Am I the only one that thinks that way?

What's in my "Oh crap bag?"

So seriously, what's in my "Oh shit bag?" No real surprises here. We have important legal documents and a few small irreplaceable "family" items. Not a lot (it's a small bag), but those are things that we would need should we lose the house and not be able to recover anything.

I've thought about expanding my "bag" and instead, have an "Oh crap van." That way, I could also load up all my family (and NAVY) photo albums, my mom & dad's "pass-me-downs," my computer(s), hard drives, backups, and a few more clothes. But then, with the hurricane season as it is—that is, hit or miss—I'd be spending a lot of time loading and unloading. Sheesh.

And of course, I need to remember to carry that one other item… our "new" alarm clock we fondly call Moosche, our cat. We can't forget him. I'd miss reveille without him.

(47) Memorable "Moments," Part 1
The Flight Deck

We all have those "moments" we remember from time to time. It may be a memory from when you were a child, it may be the time your child was born, it may be a special moment with your spouse, or it may be something from your work. It's also not unusual for one of your senses (you know… the "touch, sight, hearing, smell, and taste" senses) to trigger one of those moments or memories.

In my short four-year Navy career, I spent 2½ of those years ('75-77) aboard *USS Kitty Hawk*, an aircraft carrier based out of San Diego. I made one Westpac cruise ('75) and spent a year in the Bremerton Shipyards. I wrote about some of those times in earlier blogs.

For me, those years in the Navy were some of the most challenging, demanding, yet memorable of my life. They say that working on the flight deck is the most dangerous job in the world. I've been there and done that and… tend to agree.

But I will not argue with my Army and Marine veteran friends who have faced an enemy in combat. Life on the flight deck can be dangerous, but during my time, nobody shot at us.

My hat's off to them…and as I think about it, to *all* veterans.

We in the Navy often joke about other military branches (make sure you also read my weekly, "Navy Friday Funny") and they joke about us. But when it comes down to it, I think we all understand that while we joined for various reasons, we're proud of our accomplishments and we have each other's back.

I'm at the point in my life (I'm 65 and recently "retired") where I often remember and blog about my "better times" in the Navy. It wasn't necessarily all great for my entire four years, but I certainly had more good times than not. I'd like to think that most of us did.

These are some more memorable "moments" in my journey on the flight deck.

South China Sea—Rain Shower

During Westpac '75, most of the time we flew 14 hours a day, which meant that we had to be on the flight deck for 16-18 hours at a stretch. Flight ops were often 10a 'til midnight, but since I worked in V-1 Division/Crash, we had to be on deck an hour before and stay an hour or so after. It made for a long and sometimes weary day. But like many of my shipmates often state, "I'd do it all over again."

Kitty Hawk's overseas home port (forward deployment) was the Philippine Islands ("PI"). Since it was mid-summer and not far from the equator, we'd often encountered rain storms in the South China Sea. Many times during the cruise, the ship observed "water hours," which usually occurred when the ship was at sea for more than two weeks at a time.

That meant the ship's engineers turned off the hot water. There were times when we (working on the flight deck) didn't make it down to the berthing compartment in time to take a hot shower after a long day of flight ops. And not everyone wanted to shower in ice-cold water. It made for an interesting "aroma" in the berthing compartment.

It was during those times—and I might add, in between flight ops when the ship would be smack dab in the middle of a good rainstorm—we'd strip off our jerseys (Yellow or Red for V-1 Division) and "wash" ourselves right there on the flight deck.

No, we didn't have a bar of soap (though there was that one

guy one time…), but in that hot and humid climate, that cool rainwater felt so good.

During the summer here in Florida, I often "remember" those *Kitty Hawk* rainy days on the flight deck. It usually happens when I walk out of the grocery store with bags of groceries and the "bottom falls out" and my truck is parked the farthest away from the door because I wanted to get a little more exercise that day.

My mom used to call these summertime showers, a "frog strangler" rainstorm. It takes me back to that time on the flight deck.

Sea of Japan—Freezing Cold

On the opposite end of the temperature spectrum, also during Westpac '75, we cruised through the Sea of Japan (after our port call in Yokosuka, Japan). It was quite different from the South China Sea, weather wise. When I ventured out onto the flight deck, I wore my thermal long johns, red Crash jersey, wool sweater, Navy foul weather jacket, rain gear, and my aviator knit cap. Even with all that warm bundling, I was still freezing cold.

We worked long shifts, often flying around the clock, so that didn't help. More than once, I'd stumble in from the flight deck to try to warm up with a cup of steaming hot coffee.

Back then, we didn't have a "Mr. Coffee" coffee maker—*not* for 80 of us in V-1. We made it the old fashion way, too, in a 100-cup coffee urn. Hopefully, the guy who made the coffee used the "fresh water" spigot and not the "salt water."

And let me tell you, many times I'd come in off the flight deck and have to tilt that coffee urn to get the bottom dregs, complete with burnt coffee grounds. But in that cold weather, even *that* coffee tasted wonderful.

These days, I make coffee a *little* differently than I did back then. I'm not a coffee "snob" by any means, but I use a grinder for fresh grounds, and watch as the coffee maker "drips" the hot water over the grounds, wanting it to be done sooner rather than later, but I wait… patiently.

When it's ready, I pour a cup, taking in the pleasing aroma of freshly brewed Colombian coffee, and then close my eyes to take that first sip and I remember… how god-awful that Navy coffee with spent grounds tasted and thankfully, I don't have to drink *that* crap anymore.

Falling Asleep on the Crash Tractor

Crash had one firetruck on the flight deck and two crash "tractors." They were the same type as the aircraft tow tractors, but fitted with an open cage compartment in the back to hold the PKP firefighting retardant and a seat for the suited up crash crew member to sit.

They were painted white as opposed to Navy traffic yellow. All were mobile units and during flight ops, always idling and standing by in case… well, you know, a crash or a fire.

During recovery operations, one of the crash tractors would be "stationed" amidships alongside the "foul line." That's a straight two-inch wide alternating red & white line that extends from the angle of the ship all the way down to the fantail (the back of the ship). That's a line that you "DO NOT CROSS" during recovery operations. Doing so would put you in the middle of the landing zone, and that's a place you do *not* want to be.

Nighttime operations would be cooler, especially when we were up north in the Sea of Japan. Between the long hours (i.e., not much sleep) and the cold weather, it was not unusual for several V-1 Division guys to stand next to the crash tractor, trying to stay warm.

And more than once, I fell asleep leaning over the idling crash tractor—something about staying warm, a little "white noise" hum, and the vibration of the engine… next to a dozen of my closest friends.

Am I the only one who remembers that?

Bow Flying

I've written about this before, but it's one of those "moments" I don't want to forget. The flight deck is about 60 feet above the water line. The hull of the ship is shaped so it slices through the water like a sharp knife through hot butter. The wind generated by the ocean breeze sliding up the side of the ship's hull comes across the deck so that it helps to increase the wind velocity for aircraft to take off. At least… that's how I understand it.

In between flight ops, and sometimes at night, I'd often observe several fellow shipmates on the bow, leaning forward at a 45-degree angle. I say "observed," because they were right next to me, all of us leaning over the edge, the upper part of our body over the water with outstretched arms.

Yes, there was a safety net in the event someone leaned too far, but I never saw anyone fall into the nets. I know some shipmates climbed into them, and I can imagine at night what that would be like. I'm sure their "pucker factor" was a little high, that is, if they themselves weren't high.

The wind in your face and the water below made you feel you were really flying. Sort of like the movie, *Titanic*, when Leonardo DiCaprio and Kate Winslet were up front on the bow of the *Titanic*—you know the scene.

It was *almost* like that. Except… we didn't have anyone on board that looked like Kate Winslet, or Leonard DiCaprio for that matter.

Still, when I'm walking outside and a stiff breeze hits me head on, I often close my eyes, tilt my head back, and I remember… those days of bow flying aboard *Kitty Hawk*.

F-14 Tomcat in Full Afterburner at Night

Working on the flight deck at night is obviously more hazardous than daytime. But once you've been out on the deck several times, you get used to the rhythm of the action. It's still dangerous, but because you do many of the same things day in and day out, you know where to be and when and more importantly, where *not* to be.

One thing I remember vividly is standing near an aircraft, ready for launch. On *Kitty Hawk*, there were four catapults, numbered 1 through 4, starting on the starboard (right) bow. Cats #1 and #2 were the "bow" cats; cats #3 and #4 were the "waist" cats.

The waist is amidships and the widest part of the flight deck. I liked that spot the best at night. That way, I could stand about 20 feet away from an F-14 Tomcat (yeah, the same one Tom Cruise "flew" in the movie, *Top Gun*), and watch as the Yellow Shirt director "handed" the pilot over to the "Shooter," ready for launch.

It was loud enough to rattle your safety goggles, but would soon get much louder when the Shooter signaled the pilot to go to full afterburner. THAT is quite a sight—that much power, waiting... waiting... waiting... waiting until the Shooter signaled for the aircraft to launch.

Then, with the push of a button by the V-2 Division launch operator, the steam catapult launches the Tomcat off the deck, still in full afterburner. We watch as he flies straight for a bit, then suddenly, turns vertical in a steep climb.

All of this excitement happens within a matter of seconds. And then... we get to do it all over again in less than a minute with the next jet taxiing up to the catapult. Wow... just wow!

And to think, I was a part of that awesomeness.

Sunrises & Sunsets

How can I not mention sunrises and sunsets seen from atop the flight deck while out in the open ocean? Breathtaking... spectacular... awesome. Simple words cannot truly describe the feeling one has when experiencing the sun rising, peeking over the horizon with just the right amount of clouds and mesmerizing colors to give it that "somebody bigger than us is responsible for this" feeling.

The scene is so peaceful, it almost feels like you're completely outside your body as you try to take it all in, even though there's 200 of your closest friends on the flight deck with you. Some, like me, were "oohing and awing" internally to themselves. Some *older* salts—who won't admit it out loud—can attest to those days. And I know... *I know* when they're on the beach these days and look out over the horizon to watch the sun rise... Yeah, they remember, too.

These are several of the flight deck "moments" that I often remember. They're the happier moments I've experienced along my journey. Sometimes, though, the triggers I experience are not as happy. But I've learned to "turn" those *not so good* memories to other, happier, memories. Most of the time, it works.

(48) Not Just Another House Fire

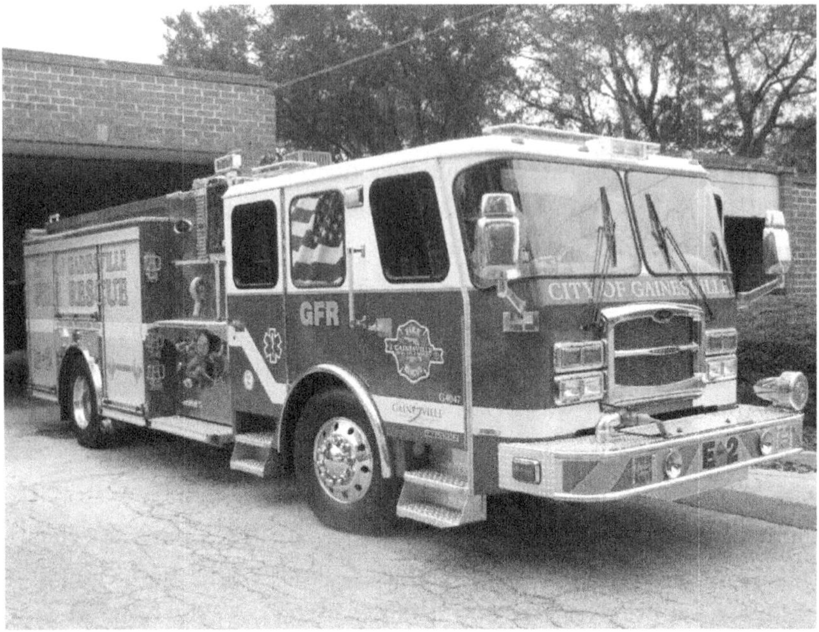

Our 24-hour shift began like most others. Muster at the station at 0730, say adios to the "B" shift firefighters, and take over duties at the City of Gainesville Fire Station #4. This was a day, just like any other day, in the fall of 1979. The Florida Gators had just lost their first game of the season… on their way to a 0-10-1 losing season. Ouch!

We go through our morning rituals: check the fire truck over and make sure all my firefighting gear is nearby. Boots & pants (bunched around the boots to quickly jump into and pull up), gloves tucked in my coat pocket, my 6' coiled rope in the other pocket, and my helmet—all hanging on the back of the fire engine. I'm good and ready to go in case of an emergency.

It's a Tuesday—*our* day for lawn maintenance. Station 4 was known for having one of the best-kept, city-owned landscaped properties, thanks primarily to Captain "Pop" Booth, Station 4 chief. Even though he worked "B" shift, we all enjoyed the same pride of keeping the grounds up to his standards.

The four of us agree on vittles for lunch and dinner, then climb onto the fire truck and drive a half-mile to Publix to purchase the day's groceries. Gordon, Engine #4 driver for "C" shift, is cooking today. I'm still new at Station 4 and have yet to "make my mark" in the kitchen. I don't mind—Gordon is an excellent cook. Herb-baked chicken, seasoned rice, green beans, and Gordon's biscuits. I won't go hungry today.

When I left the U.S. Navy in July 1977, after a brief 4-year "career," I started back to school at Santa Fe Community College. I'm on my way to becoming an Electronics Engineer, though I've still got a couple of years left.

Even though I've got the GI Bill, I needed a full-time job to help pay for college & living expenses. Being a paid city firefighter ($9,600 a year), working one day on and two days off, is great and allows me to attend school full-time. I hope to graduate with an AS degree from Santa Fe next year and transfer to the University of Florida. My GPA is currently 3.6. I like my classes.

I've been a firefighter with the city for two years and worked at all seven city fire stations, including the Gainesville Airport. It wasn't the same as NAS Agana, Guam or *Kitty Hawk*, but I still loved working out there in the boonies, watching aircraft take off and land. The best thing? We didn't have to sit in the truck for two 4-hour watches on the "Alert Spot" like in Guam.

But I digress…

After lunch, we begin our afternoon training, but the station alarm goes off. We head to the truck. I'm riding the tail gate, Kenny is the #1 hose man, riding just behind the cab, Gordon is driving, and Lt Ray is in charge. The address comes across as we mount the fire engine… 2231, about four miles away.

That address…. it sounds… familiar.

We are out of the station heading toward the house fire, lights flashing and siren blaring. My heart is racing—it always does when we go into the unknown.

About a minute out, I realize that the address is Becky's new house; I had just helped her move in two weeks earlier. I'm screaming at the top of my lungs to Kenny, "That's Becky's house! That's Becky's house!" Becky and I had been dating for a year and she'd been to the fire house several times and knew the crew.

Crap, we can't go fast enough. Kenny knocks on the back window and lets Lt Ray know what I just yelled to him. I can see Lt Ray tell Gordon, who glances in the rear-view mirror. I feel the diesel engine move a little faster, but it's a diesel… we still can't get there fast enough.

This is the woman I'm going to marry in the not too distant future.

As we get closer, I can smell the fire before I see any smoke—it's an innate ability we firefighters have. Strange smell, though; a mixture of wood & trash and something else… gasoline? What the?

We make a turn left onto the street and Gordon stops to let me jump off. One of my jobs on the tail gate is to connect a 2½" hose to the fire hydrant to feed water to the truck. There's one at the entrance to the street.

I grab the hose off the back of the truck, wrap it around the hydrant and yell to Gordon, "Go!" He speeds down the road, hose feeding off the truck, carefully maneuvering between the cars parked on either side. It looks like he's stopping a little short of Becky's house, but I can't tell… it's too far away. Besides, I'm busy hooking up the hose.

I uncap the hydrant, attach the hose, and straighten it out, waiting for Gordon's signal. "Thumbs up"—he's connected the other end of the hose to the truck and is ready. I open the hydrant and the hose quickly fills with water, expanding out even more. I learned a long time ago—the *hard* way—to *not* stand over the hose when opening the valve.

With the water hydrant secure, I run down the road—it's only about a hundred yards, but it seems like it takes forever, especially running in my fireman's boots. When I get to the fire truck, it's stopped at 2231. Wait… Becky lives two houses down at 2121. Whew, a mixture of adrenaline and relief.

I say a quick word of thanks—a guilty thanks, that is.

I quickly assess the situation—black smoke pours out of the top of the closed garage door; the front door is open, but no smoke coming out. My gut instinct tells me the fire is contained in the garage.

I see Becky on the front lawn, near the front door, kneeling beside and attending to a badly burned man lying on the ground. He's still alive and breathing, but shaking uncontrollably from shock. He's got third-degree burns on his arms and legs. Next to her is her roommate, Debbie.

They are both RNs, work at UF Shands hospital, and are well-trained and cool under pressure. I don't have time for pleasantries, but Becky instantly recognizes me. We realize we are in the same dire emergency, but attacking it from different angles. Neither of us smile, but we're both glad we're there—we've got our game face on.

Kenny's already got the 1½" hose, unfurled and with Lt Ray's help, has it straightened out as I arrive. All three of us have donned our oxygen tanks and masks, ready for anything. I know from experience, my air tank will last about 15 minutes with labored breathing.

"You ready?" Kenny asks me and Lt Ray. "Let's do it!" I yell back. I've got a lot of confidence in Kenny; he's a seasoned and a smart firefighter. The front door is already ajar—we don't have to break it down. I hear Lt Ray behind us, "The fire's in the garage."

Kenny leads the way and as we cautiously move through the living room. The smoke is so thick, neither of us can see more than a few feet in front of us. Kenny is low and makes his way into the kitchen and looks back—I'm two feet behind him, also staying low.

"When I open the door to the garage, we'll probably have a flash over. Be ready." I nod my head, acknowledging the anticipated danger. I was an aircraft firefighter for four years in the Navy, both on land and aboard an aircraft carrier. I've also fought several fires in my short two years at the fire department. I know what I'm doing… or so I thought.

Kenny is squatting down, as am I. We're both ready. Lt Ray isn't far behind, but he's letting us knock the fire down before he gets too far in. Kenny opens the door and as promised, hot orange flames shoot out above our heads, following the route of the ceiling. It's not too bad, but if I was standing at my full 6'4" frame, I'd need immediate medical attention.

Kenny's got the nozzle open wide, full spray to shield us from the fire's blast furnace. I can feel it on my face, but I tilt my head down a bit so my helmet brim will help reflect the heat. I can still see, but barely. Kenny's sweeping the hose back and forth, inching his way into the garage. This fire's not too bad. We should have it out in a matter of minutes.

The outside garage door is still down, but there's another fire crew outside, trying to pry it open to let out the smoke. We all work well together. I think this crew is from Station #5, off 13th Street, close to the Gainesville Mall. They're our backup on this scene.

I hear a faint bark. What's that sound? "Arf-Arf." Holy crap, there's a dog in here. I immediately jump up, ready to spring forward to find and rescue him. I'm a dog guy; I've had dogs my entire life and they are special to me. He's rustling around somewhere, but I just can't see him. I can sense the fear in his bark. I just hope he's not been burnt.

I start to move away from Kenny, but he grabs me and with one strong yank, I'm back on my knees. "There's a dog in here," I say. "There's a gas can over to the left, stay behind me."

Within a matter of seconds, the gas can erupts, spewing flames everywhere. We both lean back, avoiding the blast, and Kenny still attacks the fire. The wide spray on the nozzle still shields us from most of the heat.

I can still hear the poor dog. I'm afraid he won't make it, but I'm thankful for Kenny. He may have just saved my life. There's no telling where I'd be today had I gone forward to rescue the dog and that 5-gallon gas can blew up in my face.

We're safe now. It only takes us a few more minutes to put the rest of the fire out.

I can't help but think of the poor man and his poor dog. I don't know if either of them will make it.

Within a few more minutes, the other fire crew pries the garage door open, and smoke pours out, clearing the space.

The ambulance is outside, taking the man away on a stretcher. He's heading for Shands' Burn Unit, which was (and still is) one of the best in the country. I pray he'll be okay.

One of the neighbors had wrapped the puppy in a moist towel and taken it to a nearby veterinarian. Someone said that when the crew from Station 5 was working to pry open the garage door, the little puppy scurried out and collapsed on the driveway.

Lt Ray tells us that apparently the man who owns the house was working on his car in the garage with the garage door down (why, I have no idea). Something sparked, and while he tried to put out the fire, it got bigger and he got burnt… badly. He managed to crawl out the front door and a neighbor, seeing the flames and hearing his cries for help, dialed 911.

Becky and Debbie live two houses down and heard the commotion. Becky normally works as an ICU nurse at UF Shands—the late shift. For some reason, that night she "floated" to another unit… the Burn Unit… and assigned to the same burn victim whom she'd attended to earlier.

Was this fate or a coincidence?

After the fire is out and the scene secured—the man lived alone with his dog—we wrap up the fire hoses and head back to the station. That was an "interesting" day, in many ways. From the time we got the call, arrived on scene, put the fire out, secured the house, and got back to the station was probably less than two hours.

It was a house fire, and while they're all different, many are the same. But this is one I'll Never Forget!

# # #

As the years have passed, I still wonder a lot of "what ifs?" about that day. I think we all tend to do that now and then. Sometimes it's healthy, sometimes it's not.

But I'm thankful that Kenny kept me from rushing into a potential fire trap. I'm thankful that the puppy was okay—he reunited with his owner when he came home from the hospital. I'm thankful that Becky & Debbie were there to help initially treat the burn victim; their quick action probably saved his life. I'm thankful the man survived with only exterior burns and was out of the hospital within a few weeks.

Most of all, I'm thankful that Becky is still part of my "journey." We married about a year and a half after that fire. We're still happily married after almost 40 years.

Both of us are still reactive to emergencies—albeit, maybe not as quickly as before. But when we come across an accident or a fire (which is thankfully rare), we both jump to see what we can do to help. I think that's the nature of first responders and medical professionals—always and forever.

God bless our firefighters and our nurses.

(49) Westpac '75: After Six Months, We're Coming Home

Kitty Hawk left the Philippines (PI) on Thursday, November 26, 1975—Thanksgiving Day—heading back to the states. I was somewhat sad, leaving a country that had been our overseas "home" for the past six months.

I had become acquainted with several favorite haunts in Olongapo City. The people were courteous and took care of us, and I'd grown accustomed to the food and "drink." I truly thought the Filipinos were a great people and felt like the U.S. presence helped them.

But what did I know? I only saw the insides of the bars and the restaurants.

I know Olongapo City was not that well off, but then Manila probably wasn't much better. The dictator, Ferdinand Marcos and his wife Imelda, ruled the country at the time we were there. It was rumored that Imelda had over 5,000 pairs of shoes. I could've told her you can only wear one pair at a time.

How can someone have so much, yet their country have so little? But then, there are so many other countries in the world with similar problems, including our own.

Down Time

On the way back, we would again stop in Pearl Harbor for a couple of days. The trip from PI to Hawaii would take about two weeks. Once again, we established the Alert 5s and once again, we launched them several times to greet the Russians. We also spotted several Russian trawlers periodically cruising alongside us. We waved, they waved… hey, we're all friends here.

During these long periods with no flight ops, we found things to keep us occupied. For example, Airedales were not the only ones on the flight deck anymore. Many of those who worked below decks ventured out, and a few "regulars" jogged around the flight deck. *Kitty Hawk* did not have a gym, and only one weight room. This was an excellent opportunity for them to see how us Airdales lived, so to speak.

It was quite humorous to watch these guys, too. Not everybody knew their way around and you would inevitably see someone trip over an aircraft tie-down chain and smack into the flight deck. It probably wasn't funny to them, but was to us. I had been blown down many times by the south end of a north bound jet exhaust and I'd tripped over so many tie-down chains at night (I still have the scars on my shins to prove it), it was a break to watch other sailors get whacked.

I still had my radio show ("The Country Chet Adkins Show") from 0600 to 0800. Several of us onboard DJs had our own *Kitty Hawk* theme song. I started playing, *If We Make It Through December,* by Merle Haggard every morning on our way back. I didn't write it, mind you. I just played it and adapted it to the ship since we were going home and hoped we would make it there in December.

Tiger Cruise

When we arrived in Hawaii, I learned that the trip home from Hawaii to San Diego would be a Tiger Cruise. During this short cruise, *Kitty Hawk* sailors' sons, fathers, and brothers (no women) came aboard to sail with us back to San Diego. What a great idea—see what dad does at work.

The whole idea, of course, was to brainwash kids into joining the Navy when they got older. Actually, it was awesome, seeing dads and sons together, especially those on the flight deck because that is where I worked.

One of the catapult officers ("Shooter") would retire from the Navy when we returned and his son was on board. He introduced his ten-year-old to everyone on the flight deck—he was a cool little kid. Like his dad, he was very confident and cocky. A future "Shooter," perhaps.

When a Shooter leaves the carrier, it is customary to "launch" his flight deck boots off the catapult—another time-honored Navy tradition. It was heartwarming to watch the Shooter and his son go through the routine of "the launch," including the final salute to the boots, the hands in the air, swinging fingers, and then the launch of the boots.

The kid had such a great time mimicking his dad, and it brought back great memories of me and my dad.

Tax-free Liquor

On the way back from Hawaii, I learned *another* Navy tradition: we could order up to a gallon of liquor, tax-free. We could pick it up when we pulled into San Diego. I only ordered one bottle of bourbon. I guess I could have ordered more and sold it to someone who maxed out on theirs, but I didn't. I know several who did, though.

About 0500 the morning we were to pull into San Diego, the OOD (Officer of the Deck) passed the word to pick up our liquor orders. I didn't retrieve mine; 0' dark hundred was just a little too early for me, so I'd get it later. However, most of the guys who did order were lined up and ready ("locked & soon to be loaded!").

Of course, you were not *supposed* to drink while aboard the ship, but several sailors got pretty snockered before pulling into port. I laughed as I watched them stagger off the After brow and down the stairs. Fortunately, no one fell and broke anything.

Finally, We're Home

We pulled into San Diego about 0800 on December 15, 1975, the day *before* my 21st birthday. It was a beautiful clear and crisp morning (my mom used to call these "Champagne Days"). We'd already launched the squadron aircraft, which were on their way back to their own home bases. Squadron personnel were still aboard with us, though.

Coming through the bay and passing Point Loma on the port side and the Naval Air Station on the right, it truly felt like I was coming home. We had been on the Westpac cruise for what seemed like forever and other than Hawaii, this would be the first time in six months I had stepped foot on American soil.

When a ship comes back to home port from an overseas tour, the sailors man the rails, another formal Navy tradition. That means that everyone taking part dresses in the uniform of the day.

In this case, we wore winter blues. Sailors line up arm's length, called "Dress Right," around the edge of the flight deck, facing out. If the ship comes into port, they are usually at "Parade Rest." However, if there is a special occasion, such as passing the *Arizona* Memorial in Pearl Harbor, sailors manning the rails would come to attention at the appropriate time.

We'd been in and out of Subic Bay, PI seven times during our deployment. Prior to Westpac '75, we'd been in and out of San Diego at least a dozen times. Each time, coming back into our home port, was uneventful, at least for me.

But this time… coming home to our San Diego home port was special. I felt a lot of pride and happy to be returning to the U.S. Unfortunately for me, I had duty that first day. But that was alright, because I didn't have anyone meeting me on the pier.

Here's an example of another "in your face" moment I encountered during my short Navy "career." And which is probably why I didn't mind having duty that first day. I'd been in the Navy for 2½ years and drank in several countries during our deployment—well, actually *only* while in port. When I got back to the states, I realized the drinking age in California at the time was 21. I'm not old enough to legally drink. What kind of screwed up world was this?

But I digress…

Family Reunions

As we approached the pier, we could see hundreds of Navy wives, children, and other family members waving with signs of "Welcome Home, We Love You." Several signalmen on deck were signaling to their wives or their girlfriends (or both).

We didn't have cell phones back then; everyone with anyone was desperately looking over the crowd until they spotted a familiar face. Then you could see the joy in their eyes and the frantic waving, "I'm here, I'm here." It was very touching and emotional.

The new fathers disembarked first, and greeted their wives and their new babies with open arms and tears—lots of joyful tears. Even some of the older salts teared up. Even though I was still only twenty years old without a wife or girlfriend, it was a touching scene. These reunions are very heart warming.

While a few of the wives had flown to the Philippines, Hong Kong, or Japan for a brief visit with their husbands, this was the first time many of them had seen each other in six months. It was truly wonderful to see so much love and affection in one place.

Lessons Learned

So now that I was a worldly man, what wonderful things did I come away with from the Westpac '75 cruise? I had faced several close calls on the flight deck. I had seen death, and it was not pretty. But with all of that time spent during flight ops, I gained a higher respect for life.

Sure, you can grab ass now and then, but you also realize that your life can be yanked away from you in a second. It is hard to explain the confidence I had gained during the cruise. It is almost like each new duty station in my Navy career was a building block, one on top of the other. Or as I commonly say, "Life is a journey, one step followed by another."

Besides Guam which is really a U.S. territory, this was the first time I had been outside the United States. We went to three different countries: the Philippines, Hong Kong, and Japan. Each had their own cultures, their own foods, and their own traditions.

While I did not really tour PI that much, I definitely took advantage of touring Hong Kong and Japan. I think all in all, I came away with a greater appreciation for the United States. Even at twenty years old, I still had my own opinions about our country, what I liked and what I didn't like.

But as my dad (a World War II veteran) often told me, "There is no better place in the world to live than the United States."

You know what? From an ol' Florida boy, he was right.

(50) Where Has the Time Gone? It's all Relative…

I started this weekly blog, "A Veteran's Journey," last year on Veteran's Day, November 11, 2019. Today marks my 50th post. Wow—where has the time gone?

Most of the time, I post on Tuesday mornings. Periodically, I'll throw in an extra blog, depending on the day and the "occasion."

I normally write about my Navy days (1973-1977), but sometimes I'll write about my dad's World War II experiences. He fought in Europe with Patton's Third Army (80th Infantry Division). I wrote a book based on his diary: ***You Can't Get Much Closer***

***Than This: Combat with the 80th 'Blue Ridge' Division in World War II Europe*.**

At this point in my life, blogging is a way for me to continue writing and to keep my sanity.

I've received lots of comments (mostly positive, thank you), and because of these comments, several readers have reconnected with other shipmates and veterans from their past. That's been a surprising reward—that and I, myself, connecting with other shipmates through social media. Yes, there is *some* good in social media.

And oh yeah. I also blog a "Navy Friday Funny"—something to end the week on a lighter note.

# # #

"Freedom…"

There was *that* time in the late 60s & early 70s when I was a teenager and I couldn't wait to leave home. Then when I left, I quickly realized what it meant to be out on my own: responsible, accountable, and… a bit anxious.

Yeah, you remember *those* first few months of "freedom," don't you?

I left home after graduating from Gainesville High School (Class of '72—Go Canes!). I was 17 years old when I "first" went off to college—Florida State University, about a 2-hour drive from my hometown, Gainesville. It was my first time away from home and while I *thought* I was ready, I found out that I really wasn't.

I lasted for three quarters. My G.P.A. was a solid 0.7 (no, that's *not* a typo).

I was into anything that *wasn't* school-related, including scuba diving. Yes, FSU had a scuba diving course as part of its Physical Education program (my one and only "A"). In my short freshman year, my declared "major" changed from Biology (why not?) to Marine Biology (scuba diving-related) to Physical Education (much easier).

But with a 0.7 GPA, FSU sent a letter along with my grades, "Dismissed for academic reasons."

"Saved" by the Navy

I was still living in Tallahassee when my parents got the same letter the same day… *that* was an interesting telephone conversation.

Dad: "Son. We got your grades in the mail today."
Me: "Yeah. I didn't do so well."
Dad: "We probably need to talk about it."
Me: "I joined the Navy."
Mom & Dad: "What???"

As I've looked back many times over the years, I can easily claim that getting kicked out of FSU was one of the best things that ever happened to me. It jump-started me into maturity *and* to join the Navy.

Anyone else have that type of experience? If *you're* a veteran, why did you join?

After flunking out of FSU, I enlisted in the United States Navy, signing on the dotted line ("Press hard, the third copy is yours.") for a four-year commitment. I chose my rating to be an Aviation Boatswain's Mate-Handler. My two duty stations were NAS Agana, Guam and *USS Kitty Hawk*, CV-63 (an aircraft carrier).

I've blogged many times about both duty stations.

Where has the time gone?

But now that I'm retired (I'm still trying to figure out what I want to do when I grow up), I look back on my various "careers" (or "journeys," as I like to call them), and wonder… *where has the time gone?*

Some things I seem to remember with acute accuracy. But there are other times when my wife has to correct my "acute memory."

Does anyone else experience that?

Relativity-ly speaking…

There's a line in the movie, *Deep Blue Sea* (1999), where "Preacher" (the cook, LL Cool J) explains to the physicist with two PhDs, "Tom Scoggins" (Michael Rapaport) *his* own theory of relativity.

"Einstein's theory of relativity. Grab hold of a hot pan, a second can seem like an hour. Put your hands on a hot woman, an hour can seem like a second. It's all relative."

I've had similar experiences many times. I'm sure you have, too.

During our 4-hour shift on the "Alert Spot" at NAS Agana, Guam, I had to constantly look for things to keep me awake and alert, especially while on the 1200-0400 "O' Dark Hundred" watch.

There were 4 or 5 of us in the crash truck parked between two active parallel runways. NAS Agana, Guam also served as the airport for commercial airlines. Sometimes, we had some interesting & philosophical conversations, sometimes I'd read, and sometimes we played board games (though books, cards, & games were "not allowed" while on the Alert Spot <grin>).

But those four hours always seemed to drag. Sure, we could get out of the truck and walk around, but we had to remain close by in case of… you know… an aircraft crash. Being out on the Alert Spot meant we'd be the first to respond *and* the first to arrive.

Then there were those times when we had an actual aircraft emergency. I'm thankful that I *never* experienced an aircraft crash during my 15 months on Guam ('73-75). But I know others that did.

We certainly had our share of aircraft emergencies, both those that were radioed in ahead of time as well as spotting an aircraft land so hard it blew out a tire (that was fun).

But during *those* types of emergencies, time flew by so fast that hours seemed like minutes.

Relativity.

The Next Journey(s)

Fast forward to a few years *after* the Navy, out of college, married and now a working stiff. I know I'm not alone here.

You're building your career *and* in a new place (for us, Seattle) and loving the new environment (work and Pacific Northwest). Time flies by for both you and your wife.

Fast forward again to when your child (or in our case, kids) are born. You're now working "split shifts."

Yes, I can still remember those "working" nights after a full day of work, staying up with the young'uns to help calm them down after an upset tummy or a bad dream. Some nights seemed to drag on so long that I wondered if morning would ever show up.

Then they start school, then sports, then music, scouts, and other non-school-related activities. Time is flying by so fast. You learn the magic of being in multiple places at the same time and hopefully, keep track of the kids' (and your own) schedules.

Then, the kids graduate from high school, move out of the house, and head off to college. That was… tough.

Our daughter is the oldest and the first "out the door." She wanted to experience a small, liberal arts college outside of sunny & warm Florida and ended up in the frozen tundra of St. Paul, Minnesota (Macalester College).

Our son is two years younger and attended a private college here in Florida (Full Sail University).

Somewhere down the line (where did the time go?), they both graduated and are now full-time working professionals. She's a professor at Case Western University (Classical Studies) in Cleveland, and he's a UX/UI (User Experience/User Interface; I had to look that one up) game designer in San Diego.

Where did the time go? It seems like *only* yesterday…

Death is Part of Life's Journey

My parents left this worldly place way too young. My dad fought hard for a year before he succumbed to cancer at age 68. I thank the good Lord that he was never in pain and that we could be here in Gainesville with him during his last year (1989). He was my hero.

Several years after my dad died, my mom also battled cancer for a year. She was only 72. I thank the good Lord that she was never in pain and that we were here in Gainesville with her during that time (1998). She was also my rock.

But during those days of visiting in the hospital and spending time with them (together and individually) at home, there were some days when I'd sit with them and… *minutes seemed like hours.*

That was many years ago, but I can remember a lot of those special "moments" we had together during their last days on earth.

I still talk to them almost every day. I know they hear me. I've also learned to listen, and honestly, now and then, I know that they "speak" to me.

Anyone else?

Ending on a lighter note...

My wife recently found a new "definition" of life as you get older. She likened it to a roll of toilet paper. Stay with me here...

"When you first break out a new roll of toilet paper, it unravels slowly. But as you near the end of the roll, it seems to unravel much faster. Sort of like how time seems to fly when you get older."

Where has the time gone? It's all relative...

LIST OF FIGURES

(All photos copyright Andy Adkins unless otherwise stated)

23. Japan (1975), Westpac '75; Andy Adkins.
24. Hurry Up & Wait Coffee Mug; Andy Adkins.
25. VFA-131 Wildcats F/A-18 Super Hornet launches from *USS Dwight D. Eisenhower* (CVN 69) at sunset (Oct. 1, 2020); Photo credit: MC3 Cameron Pinske.
26. Andy Adkins in '54 Willy's Jeep (1972); Photo credit: Anne Adkins Berkey.
27. *USS Kitty Hawk* Manning the Rails; Otaru, Japan; Photo credit: MC3 Jarod Hodge.
28. Nurenberg, Germany (April 1945); Photo credit: A.Z. Adkins, Jr.
29. Lt. A.Z. Adkins, Jr.; Austria (May 7, 1945).
30. Andy Adkins; Bremerton Shipyards (1976).
31. *USS Kitty Hawk* Decommission Ceremony, Bremerton, Washington (2009); Andy Adkins.
32. Rest in Peace — Photographer unknown.
33. VF-213 F-14 Tomcat afterburners aglow after being launched from *USS Kitty Hawk* on six-month deployment to the Western Pacific and Arabian Sea; Photo credit: PO3 Mark Hamilton.
34. Lt. Rufus K. Broadaway (1943); Becky Broadaway Adkins.
35. NAS Agana, Guam Crash Crew (1974); Andy Adkins.
36. An Anti-submarine Squadron (HS-8) "Eight Ballers" helo hovers near *USS Constellation* (CV 64); Photo credit: PH2 Heuser, USN.
37. *USS Arizona* Memorial, Oahu, Hawaii, National Park Service.
38. NAS Agana Guam (1974); Andy Adkins.
39. *USS Kitty Hawk* Radio Station KRAL (1975); Westpac '75 Cruise Book.
40. Standard Navy Issue Meal Tray.
41. *USS Kitty Hawk* Battle Group in Indian Ocean; PH3 Brewer/PH3 Hensley.
42. Mount Rainier, Washington; National Park Service.
43. Hong Kong-Aberdeen, ABH3 Steve Cummings (1975); Westpac '75; Andy Adkins.
44. Guam — Guam Office of Public Accountability.
45. Sgt. Ralph Freeman, Lt. A.Z. Adkins, Jr.; Argentan, France

(August 1944); A.Z. Adkins, Jr.

46. Oh Sh*t Bag.
47. *USS Kitty Hawk*; U.S. Navy photo.
48. Gainesville Fire Department — City of Gainesville, Florida.
49. *USS Kitty Hawk*, Subic Bay, Philippines (1981); PHC Lawrence Foster.
50. ***You Can't Get Much Closer Than This: Combat with the 80th 'Blue Ridge' Division in World War II Europe***; Andy Adkins.

ABOUT THE AUTHOR

I'm actually in my fifth (or is it sixth?) career. I spent four years in the U.S. Navy during the 70s, most of which was on the flight deck aboard *USS Kitty Hawk* (CV 63) as a Crash & Salvage firefighter and a Yellow Shirt aircraft director. After my honorable discharge, I became a firefighter with the City of Gainesville, Florida.

By education and training, I have two degrees in electronics engineering (BSEE '82, ME '89) from the University of Florida and spent five years designing and developing microprocessor-based computer systems in the 80s.

I then turned to consulting, working with lawyers all over the country as an independent legal technology consultant. My father was an attorney, so that helped open a few initial doors. Plus, back in the late 80s and early 90s, law firms were beginning to incorporate technology into their practice. I was in the right place at the right time.

The University of Florida Levin College of Law provided an opportunity to establish the Legal Technology Institute (LTI), where I continued providing consulting services to the legal profession. During my tenure at UFLaw, I also served as the IT director and an adjunct professor, creating and co-teaching Law Practice Management to 2Ls and 3Ls. I left UFLaw in June 2010 with the intention of "slowing down." I privatized LTI and continued to consult the legal profession.

Just when I thought I was "slowing down," a previous client contacted me, asking if I would join him as the CIO for a large West Virginia law firm. We loved our time in West Virginia; though, being a lifelong "Gator," it was a little hard to say, "Go 'Eers." After four years, we moved back to Florida, intending to retire.

However, a previous employer called, asking if I would join him in a new software venture. I spent a couple of years working with a great group of folks.

Now, I'm "retired" (again)… I hope I'm slowing down.

Visit my website and follow by blog at

www.azadkinsiii.com
www.azadkinsiii.com/blog

Made in the USA
Monee, IL
04 November 2026

67379696R00154